Oil & Gas and the Texas Railroad Commission: Lessons for Regulating a Free Society

Dr. Mark A. Miller

"Aside from the cowboy, nothing is more iconically Texan than pump jacks and oil derricks. Yet few Texans know that their Railroad Commission regulates oil and gas, not railroads. My intention in writing this book is to help Texans become more knowledgeable about an obscure yet vital agency. Voters can hold their Texas Railroad Commission accountable when they elect at least one of its leaders every two years."

- Dr. Mark A. Miller

Cover design by Denise Lott Luckey

Published in the United States by Dr. Mark A. Miller

Miller, Mark A., 1951 –

CONTENTS

CHAPTER TWO

CHAPTER THREE

PREFACE

In November 2014 I was on the general election ballot for the office of Texas Railroad Commissioner. Running for public office was challenging, interesting, and fun. But it was also discouraging to discover just how little Texas voters know about the State's most important regulatory agency – an agency whose primary role is to regulate the oil and gas industry.

I learned a lot from the campaign. As a former University of Texas faculty member and 43-year experienced petroleum engineer, I already knew much about oil and gas and the industry that explores for, develops, and produces these critical natural resources. But there was much more to be learned about the Railroad Commission and the many oil and gas regulatory issues facing Texas. Many of these issues sprung out of the most recent oil and gas boom, enabled by hydraulic fracturing (fracking) technologies. This, obviously, is a significant part of the book.

While many helped put this book together, and provided immense encouragement along the way, it is only fitting that the book be dedicated to my dear departed friend, Paul Rowe. Paul was my most ardent supporter during the 2014 campaign. His untimely passing on Election Day that year still leaves a hole in my heart. He would have enjoyed helping me with the book. I hope he likes how it turned out.

Part of me wanted this to be a scholarly work. Alas, it is not. Early on I realized that I had neither the patience nor the inclination to do so. The best I could do was to provide suggested readings that I relied upon. I'll leave the research to others, who should feel free to either support or challenge what is said within these pages.

In Texas we call it the "awl bidness"

Aside from the cowboy, nothing is more iconically Texan than pump jacks and oil derricks. Yet fewer than five percent of Texas voters are aware that for a hundred years, the Texas Railroad Commission has been the primary regulator of the State's oil and gas industry. The workings of this important agency, with its misleading name, manage to escape widespread public attention. Fortunately, the shale fracking boom is bringing increased attention to a Commission that appears more beholden to oil and gas interests than to the people of Texas.

This book was written as a challenge to Texas voters – a challenge to become more engaged in matters related to an industry so vital to the Texas economy; and to become more engaged with a State agency tasked with protecting property rights, public safety, and the environment. There are three elected Railroad Commissioners, at least one of whom is on the Texas general election ballot every two years. Texas voters have a regular and on-going opportunity, an obligation even, to pay attention to what their Commissioners are doing. It's important for Texans now and perhaps even more important for future generations.

Readers will find general information on oil and gas, as well as more detailed information on production methods such as fracking. A number of recommendations can also be found for changing regulatory policy and the Texas Railroad Commission.

Feel free to skip around in the book. Let the table of contents be your guide.

Finding pay zone

The final chapter of the book provides a general discussion of regulatory policy culminating in specific recommendations for changes in oil and gas regulations for Texas. Recommendations are based on the need for both increased transparency and regulatory reform. But they're also based on a desire to change Texas regulations to better protect personal liberties and property rights.

Widespread public trust can only be ensured by a more transparent Commission with a more appropriate name, and by abolishing the Commission's role as oil and gas industry champion. The Commission must move past its image as primarily serving the oil and gas industry to one that serves all Texans.

Complex regulations are difficult to administer and often favor the largest and most well-connected. The Commission's purview should be limited to critical activities, and a regulatory sunset review policy should be implemented. A continuous focus on reform and change is the only way that outdated and ineffective regulations can be refreshed, revitalized, and minimized.

The dominant legal position of mineral rights over surface rights has, for too long, put surface rights in an unjustly inferior position. The Commission must move to become a protector of the rights of surface owners, in addition to those of commercial entities and mineral owners.

Chapter Four contains specific recommendations for addressing problems such as human-induced earthquakes, pipeline eminent domain abuse, groundwater contamination, infrastructure damage, and local fracking bans.

INTRODUCTION

Rapid technological advances were responsible for the most recent oil and gas boom in Texas and the US. Hydraulic fracturing technologies unlocked previously non-commercial hydrocarbon-bearing shale formations. Though these technologies had been under development for decades, starting around 2005 they unleashed a torrent of oil and gas production in Texas and other places in the US. By 2014, US oil production rates had increased by nearly 70% and natural gas by almost 45%. Even though a worldwide oil glut subsequently depressed prices and reduced development activity, Texas' shale resources have not been depleted. They remain a valuable asset.

This latest boom, as with past ones, was a mixed blessing. Frenzied development brought rapid economic development, good paying jobs, increased tax revenues, and wealth to mineral rights owners and industry investors. But it also brought concerns about water usage, groundwater contamination, air quality, damaged roads, eminent domain abuse, and even earthquakes. Rapid development along with the expanding geographic footprint of shales meant that these issues impacted increasing numbers of Texans.

Fracking (more precisely *hydraulic fracturing*) has become part of our national lexicon, as well as a topic of an intense national debate. While many regale at the technological revolution that unleashed previously untapped resources, others fear that fracking is an environmental disaster in the making.

In the middle of the debate is the Texas Railroad Commission. Though established in 1891 to regulate railroads and other transportation-related activities, the agency's responsibilities were later expanded to oil and gas, increasingly significantly after the East Texas field was discovered in 1930. The Commission's railroad responsibilities were largely taken over by the federal government starting in 1984. It currently has no authority over railroads. Yet several recent bills to change the Railroad Commission's name failed in the Texas Legislature.

The goal of this book is to provide more gist for the debate. Resolution of most oil and gas issues, contrary to what those on the extremes are willing to admit, are not straightforward and often highly technical in nature. The Texas Railroad Commission is charged (among other things) with providing a regulatory framework under which commercial production of oil and gas can be done in a safe and environmentally sound manner. There have been, and continue to be, both successes and failures.

As a primary regulator of oil and gas, the Texas Railroad Commission is bound to be on the regulatory hot-seat for some time. Rapidly-evolving technologies used to produce hydrocarbons from shales mean that it will be increasingly necessary for the Commission to reform its regulatory framework. How these reforms evolve could have serious implications for oil and gas in the State, as well as in other governmental spheres.

The expected focus on reform will provide an opportunity to re-think and re-formulate what it means to regulate. We can begin that re-formulation by considering how to create a regulatory culture that protects individual liberties and property rights, promotes innovation, ensures free and prosperous commerce, and protects our commonly-shared natural resources.

This book is divided into four chapters that address its four major themes. The first chapter is a short primer on oil and gas. This chapter provides background into important technical aspects of petroleum, both in terms of its physical nature as well as the basics of the methods that are used to extract oil and gas from the earth. The intent is to provide the reader with both a comprehension and an appreciation of how we obtain the oil and gas needed to fuel our modern technological society.

This theme is expanded in Chapter Two, which addresses the physical attributes of shales and the hydraulic fracturing technologies used to produce oil and gas from them. This chapter also addresses some of the complications associated with shale production. Hopefully Chapter Two will provide the reader with an understanding of both the challenges and solutions associated with safely and effectively applying this important technology.

Chapter Three addresses the Texas Railroad Commission. Understanding the Commission's historical role in Texas oil and gas is important for engendering an appreciation of the Commission's rules and regulations, as well as the scope of its authority. This chapter concludes with a discussion of the challenges facing the Commission in regulating shale production.

Finally, Chapter Four provides a general discussion of governmental regulation, providing particular recommendations for their application in a free society. The discussion of general regulatory approaches is followed by specific recommendations for regulating oil and gas in Texas. While some of these recommendations could be immediately implemented by the Railroad Commission, others likely would require statutory changes by the Texas Legislature or local governments.

CHAPTER ONE
A Short Primer on Oil & Gas

In the spring of 1972 I was about to graduate from college with a 3.5 grade point average from a prestigious science and engineering college in Southern California. Unfortunately it was a bad year for engineering graduates. The Vietnam War was winding down (though the draft was still in effect) and defense spending was declining. Newspapers ran stories about cab drivers with engineering PhD degrees. Things weren't panning out so well at the school's placement office. I began scouring advertisements for anything I could find. There were no word processors in 1972. I spent a great deal of time typing individual letters of interest.

One of the companies I wrote to was Getty Oil Company (of J. Paul Getty fame). Getty was headquartered in downtown Los Angeles and looking for engineers to work in its Ventura office. The job being offered looked to be both interesting and varied. And Ventura was on the beach! I thought: "I can do this for a while."

After World War II, discoveries of giant fields in the Middle East kept world oil prices at a fairly stable $3 per barrel (though inflation significantly eroded this value). During the 1960s there was little expansion of oil and gas production in the US because of massive low-cost supplies available from the Middle East. But by the early 1970s oil prices had begun to rise. OPEC (established in 1960) had replaced the Texas Railroad Commission as the monitor of spare production capacity. Excess US capacity had disappeared. Oil companies began hiring again in 1971.

Needless to say, I took that job with Getty – at a whopping salary of $975 per month. This turned out to be a higher salary than any of the Assistant Professors were paid at my college. My dad told me

I wasn't worth that much. I moved to Ventura, found a great loft apartment, and went to work in the oil patch. Life was great!

Even though I had an engineering degree, it wasn't in petroleum engineering. I knew virtually nothing about oil and gas.

Little hiring had occurred during the 1960s lull, so companies were understaffed. Enrollment in petroleum engineering programs fell during this period due to a lack of hiring opportunities, particularly in California. Getty hired me in part because they needed someone with computer skills. I had gotten computer training (self-taught and rare at that time) as part of my college education.

The job did, in fact, turn out to be something I could do for a while. It's just that "a while" turned out to be over forty years and counting. Along the way I went to graduate school, taught at the University of Texas, worked as an independent consultant, and started a small business. But I also learned (and taught) a lot about oil and gas. Who better to run for Texas Railroad Commissioner?

The remainder of this chapter is intended to provide the reader with an understanding of the role of hydrocarbon fuels in our modern society, along with a comprehension of the physical nature of oil and gas as it exists within the earth. Those with backgrounds in oil and gas might choose to skip this chapter. Others will hopefully find it educational and enlightening.

Why do we use so much oil and gas?

Prior to the industrial revolution of the late 18th and early 19th centuries, mankind's energy sources were essentially muscular (manual labor and animals) and biomass (primarily wood). The modern industrial era was in large part enabled by the availability and use of carbon-based fossil fuels, first coal and later oil and gas. Hydrocarbon-based (organic) fuels derived from plants and animals (coal, oil, and gas) have been the mainstay of civilization's energy needs since the industrial revolution. In 2012, 87% of world energy consumption was from fossil fuels (the US had a similar fraction). The remaining 13% was primarily nuclear and hydroelectric power, with only 2% being non-hydro renewable sources.

Fossil fuels are abundant, chemically stable, and energy-dense

The use of fossil fuels grew and remained widespread for several reasons. First, they are abundant. Vast amounts of living matter on the earth accumulated over eons of time. This accumulation, combined with the ever-active nature of the earth's surface, meant that large volumes of ancient bio-matter were captured in the earth's crust and transformed into fossil fuels.

Secondly, hydrocarbon compounds (those consisting mainly of carbon and hydrogen) are very chemically stable. Though these compounds might change form over long periods of time and under intense pressure and temperature beneath the earth's surface, chemical stability preserves their inherent structure and subsequent energy content.

Thirdly, fossil fuels are very energy-compact. Carbon-carbon and carbon-hydrogen chemical bonds when broken (by burning) release amazing amounts of energy. Engineers and scientists measure energy content in terms of kilojoules or BTUs (British Thermal Units). Energy measured in these units is not something most of us can readily comprehend. But let's try.

It takes around 9300 BTUs to heat and boil a gallon of water starting from room temperature. A million BTUs will boil 108 gallons of water. This amount of energy is equivalent to 293 kilowatt-hours (kWh). In 2013, the average Texas home used around 1200 kWh or 4 million BTUs per month. A 100 watt light bulb that remained lit for a month would use around 73 kWh. A kilojoule and a BTU are roughly an equivalent amount of energy.

According to the US Energy Information Agency (EIA), the amount of fossil fuels that contain a million BTUs of energy is summarized in the following table.

Fuel	Pounds of Fuel Containing One Million BTUs of Energy	Cubic Feet of Fuel Containing One Million BTUs of Energy
Wood	125	4.2
Coal	104	2.1
Oil	52	1.0
Natural Gas	41	1.5 (1)

(1) As liquefied natural gas (LNG)

A volume of one cubic foot is equivalent to 7.5 gallons.

Fossil fuels are easily converted to other useful forms of energy

A fourth point to be made about fossil fuels is the ease by which their energy content can be converted into useful work. By "ease" I mean a combination of the efficiency of energy conversion as well as the cost of constructing and operating power generation equipment. The combustion (burning) of fossil fuels is based on well-known and long-established technologies. Today's energy conversion technologies (power plants and internal combustion engines) are cost-effective, efficient, and safe.

According to the EIA, electricity generated by coal, oil, and natural gas power plants in the US currently convert, on average, around a third of the intrinsic fuel energy content into electrical energy (before transmission and usage losses). Newer technologies, such as combined-cycle gas turbine plants, can achieve conversion efficiencies as high as 60%.

And finally there is the cost advantage, as can be clearly demonstrated by the following table.

Fuel	Mid-2015 Price	Fuel Cost Cents per kWh
Coal	$55/ton [1]	2.9
Oil	$50/barrel [2]	8.8
Natural Gas	$2.80/MCF [3]	2.8

(1) 1 ton = 2000 pounds

(2) 1 barrel = 42 gallons

(3) 1 MCF = 1000 cubic feet (134 gallons) measured at atmospheric conditions. An MCF of natural gas (methane) occupies 1.5 cubic feet when liquefied (LNG)

In 2015, Texans paid, on average, around 12 cents per kWh for residential electricity.

In May 2015, over 80% of total US energy consumption was from fossil fuels. Approximately 20% of fossil fuel use was coal, 32% natural gas, and 48% oil.

The above table demonstrates why oil is not generally used as a fixed-source fuel (*i.e.*, cost). Fuels derived from oil (*e.g.*, gasoline, diesel, jet fuel), however, are easily transported by truck, rail, or pipeline. They are obviously also carried and directly used by motorized vehicles with internal combustion engines. Liquid petroleum fuels accounted for 92% of the energy used for transportation in the US in 2014. Biofuels (ethanol and biodiesel) contributed around 5%, natural gas 3%, and electricity less than 1%.

Another important fossil fuel is Liquefied Natural Gas (LNG). LNG is methane gas cooled to **negative** 260 degrees Fahrenheit. Liquefication of methane achieves a high energy density for transportation purposes. There is currently a large global trade in LNG. It is also being increasingly deployed as a transportation fuel, particularly for rail and long-distance trucking. Significant public LNG refueling capacity (principally for trucking) was under development in the US in 2015.

Ancient organic matter produces oil and gas

The primary source of oil and gas is decayed organic matter (plant and animal remains, primarily micro-organisms). Organic matter is the primary source of a class of compounds known as *hydrocarbons*, indicating that their primary constituents are hydrogen and carbon. Naturally-occurring oil and gas include a large variety of hydrocarbon compounds, including some that have other elements attached to them (*e.g.*, nitrogen).

Natural hydrocarbon compounds are chemically diverse, stable over long periods of time, and relatively non-reactive. When these compounds are burned in the presence of oxygen (from air), strong chemical bonds are broken, releasing large amounts of energy.

Oil and gas chemistry

The primary constituent in what are called *natural gases* is methane. Methane has one carbon and four hydrogen atoms. Natural gases also contain significant amounts of heavier hydrocarbon molecules such as ethane (two carbon atoms), propane (three carbon atoms), butane (four carbon atoms), and some with even longer carbon chains. Crude oils contain primarily large, long-chained hydrocarbon molecules with lesser amounts of short-chained constituents. Natural hydrocarbons occur as a continuum, ranging from the lightest gases (mostly methane) to the heaviest oils (such as found in the Canadian tar sands). Petroleum reservoirs produce both oil and gas, though in varying ratios depending on the reservoir.

Naturally-occurring oils and gases may also contain varying amounts of inorganic compounds. The most common of these are hydrogen sulfide, carbon dioxide, nitrogen, and water. They may also contain heavy metals, salts, and other compounds. Coals contain significantly more inorganic compounds (many harmful) than oil and gas.

Classification of oil and gas

Produced hydrocarbon fluids are characterized in a variety of ways. Produced oils are typically characterized in terms of their density, measured in a unit called API gravity and reported as °API. API gravity is defined such that water has a gravity of 10°API (density of 8.3 pounds per gallon). Lighter fluids (*e.g.*, oils) have higher API gravities. A typical crude oil has a gravity of 30°API (density of 7.3 pounds per gallon). Lighter (and more valuable) oils have gravities up to 60°API (density of 6.2 pounds per gallon) or more. Some tar sands contain hydrocarbons that are even denser than water (less than 10°API).

Heating value is the most useful way to classify natural gases. Pure methane has a heating value of around one million BTUs per MCF (thousand cubic feet measured at roughly ambient conditions). Heavier (and more valuable) gases have higher heating values. Almost all natural gases are processed through a facility to separate methane (for home and industrial use) from longer-chained hydrocarbons such as propane and butane.

There are a variety of terminologies used for natural gases and the liquids processed from them. A few of these are:

> **CNG (Compressed Natural Gas)**: principally methane at high pressure and ambient temperature for storage and transportation.

LPG (Liquid Petroleum Gas): propane and butane, which are gases at normal conditions but liquid under moderate pressures.

NGL (Natural Gas Liquid): hydrocarbon compounds heavier than methane that are separated from natural gases in a processing plant.

LNG (Liquefied Natural Gas): methane that has been liquefied at ultra-low temperature for storage and transportation.

GTL (Gas to Liquids): hydrocarbon liquids chemically converted from methane by Fischer-Tropsch technologies.

There's water in the subsurface, along with oil and gas

Sedimentary rocks in the subsurface are almost entirely filled with water. Subsurface waters range from fresh (drinkable) groundwater to water that is totally saturated with dissolved mineral salts. Natural brines (as well as seawater) contain more than just ordinary table salt (sodium chloride). Some of the natural constituents dissolved in subsurface brines are harmful to life. When produced, these waters must be disposed of.

Fresh water is usually considered to have a salinity less than 0.05% (500 parts per million). Seawater has a salinity of around 3.5% (35,000 parts per million). Completely saturated subsurface brines can have a salinity of 26% (260,000 parts per million) or more. The term *brackish water* refers to water that has an intermediate salinity between fresh water and seawater. Brackish waters are found at depths below the groundwater table as well as on the surface in the near seashore where river flows first intermingle with oceans.

Though much produced water is re-injected into reservoirs to enhance further production, significant volumes must also be disposed of. This is nearly always done by injection into deep geologic formations using wells specifically designed and operated for wastewater disposal.

Where is oil and gas found?

One reason that Texas is such a prolific oil and gas province is the enormous volume of porous *sedimentary rock* that underlies the State. Sedimentary rocks are formed by the deposition of material eroded from the Earth's surface and deposited in bodies of water, principally ocean floors. They exist in strata (or beds) along the edges of continents. Where Texas and Louisiana are located has been an ocean floor many times over during the last several million years. Sedimentary rocks are both source rocks for hydrocarbon generation as well as traps for oil and gas reservoirs.

The process of transforming previously-living organic matter into oil and gas is called *maturation*. Organic matter from which oil and gas are derived is most typically deposited and accumulated in seafloor sediments. Given the right geological conditions, deposited organic matter is preserved and then buried by additional overlying sedimentation. As the earth's sediments compact over time, they are buried deeper and deeper below the surface. A combination of high pressure and high temperature at depth first transforms deposited organic matter into what is called *kerogen* and then to oil and gas. This process can take millions of years.

Subsurface formations where organic matter is transformed into oil and gas are called source rocks. Oil and gas that is formed in *source rocks*, being lighter than entrained water in the sediments, slowly migrate toward the earth's surface. Some will reach the surface (*e.g.*, in seeps). But some accumulates in subsurface geologic features that capture oil and gas into *traps*. This process is called *migration*.

Oil and natural gas accumulations found within the earth's crust are commonly referred to as *reservoirs*. A common misconception is that petroleum exists in pools, suggesting large underground caverns. Only in extremely rare cases is this true. Rather, a petroleum reservoir consists of solid rock with an internal porous structure (imagine beach sand) that contains the fluids. For an oil or gas reservoir to be present, all of the following must have been available: a) a source of oil and gas, b) porous and permeable rock, and c) a trap that acts to prevent fluid movement to the surface.

What are subsurface reservoir rocks like?

There are three broad categories of sedimentary *rock types* – sandstone, limestone, and shale. Sandstones are made from silica minerals (sand). Limestones (or carbonates) are generally deposited from organic material consisting of shells or other similar materials. Shales are extremely fine-grained sedimentary rocks that can be composed of either silica or carbonates. Though shales are porous, their capacity to flow fluids (permeability) is extremely low.

Some reservoirs are what are termed *naturally fractured*. Natural reservoir fractures are caused by naturally-induced brittle rock failure. Natural fractures can exist in essentially any kind of rock. Like the pores within normal sedimentary rocks, natural fractures can provide both storage capacity as well as conduits for fluid flow.

Geologic traps keep oil and gas in the subsurface

The most common kind of reservoir traps are due to the folding and faulting (large-scale breaking) of rock structures resulting from geologic forces within the earth's crust. Folds are the most common structures formed in mountain chains. Upward arches are called *anticlines*. Troughs are called *synclines*. Anticlinal traps prevent further upward movement of hydrocarbons by virtue of their shape combined with an overlying seal or *caprock*, a layer of impermeable rock above the reservoir. Shales typically provide the caprock for these types of reservoir traps. Other more complex types of traps also exist. Identifying geologic structures that can contain oil and gas is the subject of exploration geology and geophysics.

Oil and gas reservoirs exist at high pressures and temperatures

Due to their depth of burial, all petroleum reservoirs exist in their natural state at a pressure and temperature above that at the earth's surface. For most reservoirs, the natural *hydrostatic pressure gradient* is around 0.45 pounds per square inch per foot of reservoir depth. This means that a reservoir at a depth of 10,000 feet will have a natural original pressure of 4500 pounds per square inch. One at 5000 feet would have a pressure half that, or 2250 pounds per square inch. A typical automobile tire pressure is around 35 pounds per square inch.

In some formations (many found in the Gulf Coast of Texas and Louisiana), natural reservoir pressures may significantly exceed hydrostatic pressure. Such reservoirs are called *abnormally pressured*. Natural pressure gradients for these reservoirs can be as high as 0.80 pounds per square inch per foot of depth, or more.

As a reservoir is produced, its reservoir pressure declines. Reservoir pressure decline is sometimes arrested in reservoirs with strong natural water drives, or if an injection project is initiated by a field's operator.

Natural reservoir temperature is a result of deep heat sources within the earth. Most natural temperature gradients are between one and two degrees Fahrenheit per hundred feet of reservoir depth, though higher gradients are also possible. A 10,000 foot reservoir would thus have a temperature 100 to 200 degrees hotter than the average surface temperature. Unlike reservoir pressure, reservoir temperature remains essentially unchanged as a reservoir is produced. The only major exceptions are in thermal recovery projects where reservoirs are artificially heated by injection of hot fluids such as steam.

Oil and gas is increasingly unconventional

Unconventional oil and gas refers to those resources that are not economically producible by typical (*conventional*) well drilling and completion methods. A well's *completion* is a term that refers to the mechanical condition of the well's bottomhole connection to a reservoir. Completions are designed and constructed so as to allow efficient fluid entry from a reservoir into a well.

In response to the oil shortages of the 1970s, three classes of unconventional resources were the target of new technology development. These consisted primarily of tight (low permeability) gas reservoirs, methane entrained in coal beds, and tar sands. Improvements in technology ultimately led to each of these resource classes becoming economically viable. Texas has a very significant amount of tight gas resources, but essentially no coalbed methane or tar sands.

Oil and gas from shales

More recently, the term unconventional has been extended to production from shales. Though the term *shale* is used somewhat loosely in the oil and gas industry (compared to its more scientifically-accepted definition), it has come to mean any ultra-tight (ultra-low-permeability) sedimentary rock. Two observations about shales are important for this discussion.

First, shale resources are sometimes also referred to as *resource plays*. This term means that shales are both the source and the trap (container) of producible oil and gas. Shales contain hydrocarbons that have not migrated upward into shallower conventional traps.

Since they are "self-trapping", shale resources do not require shallower trapping mechanisms that are necessary for conventional oil and gas.

The second thing to note about shales is how big a resource they represent. Shale rocks are by far the most prevalent sedimentary rocks within the earth's crust. Potential for production in both the US and worldwide is enormous, though not every shale has the precise physical conditions required to be commercially producible.

The presence of oil and gas in shales has been known for a very long time. But the low rates at which shales release oil and gas by conventional drilling and completion techniques previously made them uneconomical to produce. Industry old timers, such as me, previously referred to shales as *non-reservoir*.

Shale production was unlocked by a new technology (*hydraulic fracturing of horizontal wells*) that enabled commercial hydrocarbon production from these earth formations previously thought too poor to be producible.

How much oil and gas can be produced?

The amount of oil and/or gas that can ultimately be produced from a reservoir depends not only on the hydrocarbons originally in-place, but also on the natural *drive mechanisms* that provide the energy to move fluids to wells and ultimately to the surface. Pressure is the primary driving force to move hydrocarbons. High pressure inside a petroleum reservoir pushes fluids toward lower pressure inside wellbores. Reservoirs with drive mechanisms that sustain the highest pressures will provide the most recovery.

In many oil reservoirs, a gas phase forms in the reservoir, coming out of solution from the oil as pressure declines. This mechanism is similar to what happens when carbon dioxide is released from a soft drink. The gas phase that is formed in the reservoir expands, supplying energy to maintain reservoir pressure. Typical recovery efficiencies for *solution gas drive reservoirs* are 5 to 30%, with 20% often considered as an average. The vast majority of oil reservoirs in the world have a solution gas drive. Oil reservoirs that start out with a gas phase already present (*gas-cap drive*) can be expected to have higher recoveries of 20 and 40%.

When an oil reservoir is surrounded by a certain type of saltwater aquifer, water will naturally invade the reservoir, providing significant additional pressure support. This pressure support can result in recoveries of 35 to 75%. A common secondary recovery process called *waterflooding* is often used to artificially increase the recovery efficiency above the 20% or so expected with natural solution gas drive. Waterflooding increases recoveries to those that might be expected from natural *water drives.* Waterflooding involves the injection of water into separate injection wells, effectively creating an artificial water drive.

There are other methods, in addition to waterflooding, that are sometimes used to enhance oil recoveries. The most important *enhanced recovery* techniques involve the injection of carbon dioxide or steam. Improved enhanced recovery techniques are continually being researched, including the use of such things as surfactants (soaps) and microbes.

Most of the carbon dioxide used for enhanced recovery today comes from subsurface formations that naturally contain trapped carbon dioxide gas. There is much on-going research, however, in the potential for using human-produced carbon dioxide (*e.g.*, from electrical generation plants) for enhanced oil recovery. *Sequestration* would remove man-made carbon dioxide from the atmosphere and help reduce carbon emissions.

Unlike oil reservoirs, gas reservoirs typically have much higher recovery efficiencies. It is not unusual for a gas reservoir's ultimate recovery to be 80 to 90% of the original gas in-place.

For legal and economic purposes (*e.g.*, Securities and Exchange Commission filings), oil and gas operators are required to report what are termed *reserves*. The most common category of reserves are called *proved reserves* that provide an **estimate** of the remaining amount of oil and gas that is expected to be produced – with at least a 90% certainty. Reserves represent expected **future** production, not original or recoverable resource in the ground.

Peak oil

Readers who follow oil and gas matters know that there has been much discussion about what is called *peak oil*. Peak oil was based on a theory of M. King Hubbert, a well-known Shell geoscientist. Hubbert's peak oil curve was generated from a statistical analysis that projected expected production rates from mature oil and gas provinces, as the remaining resource base is depleted.

The US, in fact, very closely followed Hubbert's curve through 2008. US oil production peaked in 1970 at just over 10 million barrels per day. Production declined, as Hubbert predicted, until reaching 5 million barrels per day in 2008.

More recently, however, the US has experienced a significant deviation from Hubbert's peak oil curve. With the increased production unleashed from shales through hydraulic fracturing, US oil production in early 2015 had **increased** to over 9.5 million barrels per day. Many believe that the previous US peak production rate will be exceeded at some point in the near future. Hubbert's statistical analysis failed to take into account both the availability of a totally new resource base as well as dramatic improvements in production technologies.

In 2014, US oil **consumption** was just over 19 million barrels of oil per day. Worldwide oil consumption that same year was over 92 million barrels of oil per day.

Faster is better

For hydrocarbon production to be commercial, consideration must be given to both the amount available as well as the rate at which it can be produced (and sold). Though reserves represent aggregate future income, the rate at which reserves can be extracted affects their value as well. High production rates, and thus more immediate revenues, provide a short time to payout of large up-front investments, as well as funds to re-invest in further drilling and development projects or other economic opportunities.

The production rate from a field is governed by four primary factors (though there are many secondary ones): number of wells, size of the resource, reservoir pressure, and a reservoir rock's *permeability.*

Permeability is a physical property that reflects a rock's inherent ability to flow fluids. Permeability, in the US, is measured in units of *millidarcies.* This unit is named after Henry Darcy, a French engineer who is credited with formulating the equation most commonly used to describe flow through porous materials such as sedimentary rocks.

The very highest reservoir permeabilities (rare) are in excess of 1000 millidarcies (one Darcy). Conventional reservoirs typically have permeabilities in the 1 to 100 millidarcy range. Gas reservoirs are classified as tight (by Internal Revenue Service rules) if they have permeabilities less than 0.1 millidarcies. Most producing shale reservoirs have permeabilities in the range of 10 to 1000 **nanodarcies.** A nanodarcy is one millionth of a millidarcy. Even a 1000 nanodarcy "high-permeability" shale is still 1000 times less productive than a low-permeability conventional reservoir.

Production into a vertical well completion (typical for conventional reservoirs) is constrained by the relatively small area (zone thickness times the circumference of the wellbore) available for flow. Permeabilities of conventional reservoirs are sufficient, however, to enable commercial production. For unconventional reservoirs, given their low permeabilities, additional area must be created to allow for commercial production.

How to produce faster

The technology that unlocked the potential of shales combined massive hydraulically fracturing with the drilling of long horizontal wells. The oil and gas industry had already developed the technology to drill long horizontal wells (one to two miles or more) in the decades prior to significant shale production. The additional technology required was that needed to create a series of open fractures (cracks) connected to the horizontal wellbore.

In a fracked well, oil and gas flows into the surface *faces* of the fractures then travels through the fractures, into the wellbore, and up to the surface. The area available for flow in a shale well can easily be several million square feet. Compare this to a conventional vertical well that may have a flow area of only a few hundred square feet (into a perforated vertical pipe) or less.

The secret to commercial shale production has been reasonably-priced technologies that overcome the million times smaller permeability with a million time's larger area.

Many oil reservoirs and most gas reservoirs will flow naturally, requiring only a string of pipe (tubing) inside a well to contain flow to the surface. Most oil wells, however, eventually require some form of *artificial lift* to bring fluids to the surface in economic quantities. Beam pumping (also known as *horse head* or *pump jack*) units seen in oilfields are the most commonly-used method for bringing fluids to the surface. Other methods include *gas lift* (injecting gas to lighten the fluid as it travels up tubing) and downhole *electrical submersible pumps*.

After a well is drilled and completed, surface equipment must be installed to: a) separate oil, gas, and water, b) measure produced volumes, c) store produced fluids, and d) transport fluids to their final destination. Produced gases are usually processed to remove heavier hydrocarbons, vaporized water, and contaminants prior to putting them into a transportation pipeline.

Carbon emissions

No discussion of oil and gas would be complete without a discussion of carbon emissions and climate change. The issue, as most readers are probably aware, is that mankind's use of carbon-based fuels has increased the amount of carbon dioxide in the atmosphere.

When hydrocarbons react with oxygen from the air, a number of chemical species are generated, the most significant being water (H_2O) and carbon dioxide (CO_2). Essentially (though it's actually more complicated than this), hydrogen atoms become part of water molecules and carbon atoms become part of carbon dioxide.

Carbon dioxide is a greenhouse gas, meaning that large amounts in the atmosphere absorb and retain heat from the sun that might otherwise re-radiate into space. Recent increases in the earth's surface temperature have coincided with an increased use of fossil fuels. This fact, combined with climate modeling studies, have led many to conclude that, in the absence of reductions in carbon dioxide emissions, the earth is destined to become significantly warmer. Some suggest that this warming trend could produce climate changes that would be harmful to mankind.

Different fossil fuels generate different amounts of carbon dioxide. Fuels with lower hydrogen-carbon ratios generate more carbon dioxide per unit of energy produced than those with higher hydrogen-carbon ratios. An ideal fuel is elemental hydrogen, as it produces no carbon dioxide (only water) when burned. Elemental hydrogen, unfortunately, does not naturally exist in large quantities.

The following are approximate hydrogen-carbon ratios for common fuels, along with the amount of carbon dioxide emitted by each.

Fuel	Average Hydrogen-Carbon Ratio	Pounds of CO_2 Produced per Million BTUs Energy
Wood	0.1	255
Coal	0.5	210
Oil (Gasoline)	2	157
Natural Gas	4	117

What should be done about climate change?

Climate change is only indirectly an issue related to the regulation of oil and gas production. The issue invariably does come up, however, when discussing fossil fuels. A few observations are in order here. From one who is a scientifically-trained pragmatic type (and who also has much experience with computer modeling) I would offer the following observations for consideration.

- There is considerable evidence that the earth is, in fact, warming. It has been continuously warming since the Little Ice Age ended around 1850. It is important to note, however, that the earth has sequentially warmed and cooled over its entire history. Periods of natural warming and cooling are to be expected. There is some evidence, in fact, that since 1998, global temperatures may have actually plateaued and perhaps even declined slightly (though this is in dispute).

- The real issue, regardless of a hiatus, is how much of the recent warming trend is due to human activity and how much is due to natural cycles. Computer modeling by the world's most renowned climate scientists suggest the former (though their models failed to predict the recent leveling off of global temperatures). In my experience, computer model forecasts, though very useful, should never be taken as **truth**. Skepticism is a fundamental principle of science that should not be confused with denial, especially when it comes to computer modeling.

- Regardless of the degree to which human activity is impacting global temperatures, reducing the amount of carbon being released into the atmosphere would appear to be a good idea. The question, of course, is at what cost. To assume that large amounts can be withdrawn without significant economic cost and disruption is, at best, naive.

- In 2013, China emitted nearly twice the amount of carbon dioxide as the US. Together China, India, and Russia emitted over 40% of the world's carbon dioxide, compared to around 15% by the US and just over 10% by the European Union. Yes, reducing carbon emissions in the US could make a significant difference, but our ability to coerce or cajole China, India, and Russia to change their ways is unlikely. China's emissions, in particular, are expected to grow significantly because of the amount of coal they use to generate power.

- The most cost-effective way for the US to decrease carbon dioxide emissions in the short run is by increasing our use of natural gas in place of coal. The switch to natural gas, driven by low prices as a result of the shale boom, has already significantly reduced US carbon dioxide emissions since 2007. Natural gas produces around half the carbon dioxide emissions as coal for the same amount of energy. Other factors make the substitution of natural gas for coal even more attractive. These include the large amount of carbon dioxide produced when mining, transporting, and powderizing coal – as well as coal's harmful pollutants (*e.g.*, heavy metals).

- Given the projected large costs and the political realities facing the development of a worldwide consensus to pay for the enormous cost of significantly decreasing carbon emissions, a more sensible (and likely less expensive) policy might be to begin planning for how the world should adapt to climate change, as opposed to how it might be prevented. In any regard, given that climate change has been a continuous feature of the earth's history (irrespective of man-made carbon dioxide), planning for this eventuality would seem prudent.

Shales will likely continue to deliver inexpensive natural gas that will drive carbon dioxide emissions in the US even lower (at the same time massively contributing to the US economy), all without government interference. Hopefully this trend will continue.

CHAPTER TWO
To Frack or Not to Frack, That is the Question

In 2009 I was hired to consult for a company that I had previously worked for. This company had created an industrial consortium to study the performance of hydraulically-fractured wells in the Haynesville shale (East Texas and West Louisiana). Little did I know where the gig would lead.

My petroleum engineering sub-discipline is *reservoir engineering*, an area of study concerned with underground fluid flow and the long-term performance of producing wells. I was asked to help with some rather unique mathematical modeling issues that had been uncovered through studying the production performance of the Haynesville wells.

I had never previously worked on, or even looked at, the intricacies of shale well performance trends. But it did not take long for me to recognize that shale well performance characteristics were **totally** different than those seen in conventional wells. A hundred years of well performance experience from conventional reservoirs essentially needed to be re-done ... a former academic's wet dream!

I later accepted a position as Chief Technology Officer for this same company. And when it was acquired by another company, I worked with a business partner to start a small company that produced software to analyze shale well performance.

In working with shale well performance over those next five years, it became clear that the oil and gas industry was entering a new era. Not only were new technologies being deployed to previously non-commercial subsurface formations, but entirely new geologic and engineering understandings were needed in order to efficiently develop these resources. Many open questions were (and still are) on the table: how should wells be drilled and completed, how would wells interact with each other, how should reserves be estimated?

A former academic colleague of mine once asked me why operators continued drilling when there were so many unanswered questions. The answer, of course, was that they made money even when they didn't understand. Late 2014 brought a pause in shale development. I say "pause" because the resource has not gone away. Engineering and geological understanding of the resources **will** increase. Consequently, technology **will** develop further, likely becoming even cheaper to deploy.

The purpose of this chapter is two-fold. First is to provide a background into the physical processes involved in the hydraulic fracturing of shale formations. This is extremely important for understanding the technological and regulatory challenges involved. The second purpose is to provide a discussion of some of the more contentious issues surrounding fracking that have become part of the public debate. Hopefully the reader will come away with a better understanding of these issues, especially those that are dealt with in more detail in the last two chapters.

We've been fracking for a long time

The technology that enabled the shale production boom is called *massive hydraulic fracturing*. The shorthand word adopted by the oil and gas industry, *fracking*, has also been adopted by the environmental community as a pejorative. My goal in this section of the book is to provide an understanding of what fracking technology entails, with the hope that it will perhaps help lead to more rational discussions than those that have too often occupied the public arena. Hopefully readers will infer neither positive nor negative connotations with my use of the word.

Hydraulic fracturing technologies have been around for many decades. Oil and gas drillers discovered by the 1950s that it was possible to use high-pressure fluids to create open cracks (fractures) connected to wellbores. As discussed earlier, these cracks create additional area into which fluids can flow, thus significantly increasing production rates. Though the process requires costly up-front (capital) expenditures, the resulting income stream usually provides significantly higher rates of return. In most shale formations, no commercial production is possible without fracking.

The technologies required for hydraulic fracturing were greatly enhanced (and made less costly) when the industry pursued tight gas by hydraulically fracturing vertical wells. Hydraulic fracturing of vertical wells has been used for over 60 years. It is estimated that between 2011 and 2014 over 25,000 wells were drilled and fractured in the US.

How fracking is done

The physical process of hydraulic fracturing involves injecting a mixture of water, chemicals, and sand (though other materials such as ceramics are sometimes substituted for sand) into a well. Surface pumps generate high pressure at the bottom of a well to initiate cracks in the subsurface rock. Fractures propagate away from the well and become longer as fluid continues to be injected. When pumps are shut down, fractures that would normally close remain propped open by the sand (called *proppant*) that was included in the fluid mixture. The propped fracture provides not only increased surface area, but also a conduit for fluid flow into the wellbore.

At depths below around two thousand feet, fractures that are created have a vertical orientation, a consequence of the natural stresses in the earth's crust. It is not uncommon for fractures to extend a few hundred feet away from a wellbore. Fractures also grow vertically. Their vertical extent is similar to (though typically less than) their horizontal extent, and limited by geological layering above and below producing intervals.

Chemicals included in fracturing fluids are designed to reduce pumping friction, aid the transport and placement of proppant, and prevent corrosion, microbial activity, and scale formation. Though these chemicals are used in very small quantities, some are hazardous and require special handling and disposal. Companies that provide hydraulic fracturing services and materials have been actively researching alternative fracturing chemicals. Many of the alternatives being investigated are based on chemicals that have already been approved for use by the food industry.

Fracking shales

Shale production was enabled essentially by extending hydraulic fracturing of vertical wells to horizontal wells. Horizontal well drilling had already been well-developed, bringing costs down to economically-viable levels. Mile-long horizontal wells are very common, though much longer wells are also easily drilled.

Hydraulically fracturing a long horizontal well involves generating a series of multiple fractures along the wellbore. As might be expected, the process is more difficult, more expensive, and more time consuming than fracturing a vertical well.

Fracturing a horizontal well is done in *stages*, starting from the well's toe (furthest horizontal distance, *i.e.*, the deepest drilled distance) and working toward the heel (point of curved transition from horizontal to vertical). Each stage creates a small number of fractures from a short portion of the horizontal section of the well. Individual fractures are initiated at *perforation clusters*, groups of holes punched into the formation. Modern methods for creating perforations in wells (through steel casing and cement) are based on technologies similar to those used in armor-piercing artillery shells.

Though design parameters are both variable and evolving, a typical stage might have four perforation clusters spaced on the order of 50 feet apart. A 5000 foot horizontal well could require perhaps 25 stages to complete, each stage being on the order of 200 feet long. Each stage is completed by pumping a large amount of fracturing fluid down the wellbore and into the open (perforated) entry points. The fractures created at each perforation cluster are the same sort as those in vertical wells.

After fracturing fluid has been pumped into the first stage, a plug is set in the wellbore to isolate the perforation clusters in the first stage from the rest of the well. The process is repeated, stage-by-stage, until the well is finished. After all fracturing is complete, plugs in the wellbore are drilled out and the well is put on production.

The amount of water and sand required to fracture shale wells is variable. A typical well might use something on the order of five million pounds (2500 tons) of sand and five million gallons (120 thousand barrels) of water. In the early years of hydraulic fracturing development, the fluid used for fracking included chemical gels and other substances to aid the transport of proppant sand. These chemicals allowed for a maximum amount of sand to be transported with a minimum amount of fluid. More recently, a process of *slick water* fracking has mostly replaced gel fracking. "Slick water" refers to chemicals that are added to reduce flowing friction so that fluids can be injected at higher rates. These chemicals are similar to those used to increase water flow through fire hoses. The use of slick water requires fewer chemicals than gel fracking, but larger amounts of water.

The geography problem

One of the largest issues that has plagued the development of shale resources is its extremely large geographic footprint compared to conventional resources. The Eagle Ford shale in Texas, for example, is roughly 50 miles wide by 400 miles long, extending over 26 counties that stretch from north of Houston to the Mexican border (and beyond). The Marcellus shale in the northeast underlies 95,000 square miles from New York to Virginia.

Historically, oil and gas activity has largely been confined to sparsely inhabited regions of Texas (*e.g.*, Permian Basin, East Texas). The granddaddy of shales, the Barnett in North Texas, however, abutted inhabited areas from the beginning of its development. As communities grew and Barnett shale production extent expanded, more and more Texans subsequently came in contact with shale development.

Though the first well was drilled in the Barnett in 1981, it wasn't until 2003 that drilling activity began to accelerate, as operators began applying horizontal drilling techniques. By 2015 there were nearly 20,000 wells producing from the Barnett, covering 5000 square miles over 25 counties.

The evolution of the Barnett provides lessons for what can be expected across the rest of the State as shale development continues. Drilling in the newest shale plays (*e.g.*, Eagle Ford in South Texas and Permian in West Texas) has largely taken place in sparsely inhabited areas of the State where there was already prior significant oil and gas production. Even still, many small communities in these areas have already felt some of the negative impacts of fracking operations. Barnett shale development suggests that increasing numbers of Texas residents are likely to be affected in the future.

In the early stages of a field's development, wells are drilled fairly far apart. Early drilling is done primarily to satisfy requirements to hold leases signed between operating companies and mineral rights owners. But as a play matures, considerable *infill drilling* occurs. It is not unusual for there to ultimately be eight mile-long shale wells drilled inside a square mile (called a *section*).

This sounds a lot worse than it is. Generally eight or even sixteen or more wells can be drilled from a single pad. If possible (given regulations, land positions, etc.) two square miles of land can often be developed from a single relatively small pad site, thus greatly reducing the impacted surface footprint.

But even though shales may be developed with a modest footprint, as development moves near inhabited areas, surface property owners are negatively impacted. Late-stage development of the Barnett shale is a case in point. The resource has been developed over a very long time. Yet there are still opportunities for commercial production by low-cost producers, sometimes in the less commercially-attractive areas of a play. Unfortunately, some of these areas are close to, or sometimes even inside of, established communities.

Once shale wells have been drilled and completed, the disturbance to nearby residents is relatively small. Some surface equipment is required to process, measure, store, and transport produced fluids. These facilities, however, can generally be located so as to be relatively unobtrusive to homes and businesses.

Disturbances in residential areas are more severe during well drilling and completion operations. Drilling and completing a single well often takes several days, sometimes a few weeks. The impact on those living near the operation includes heavy traffic, noise, night-time lighting, and smells. And even though the process is temporary, its impact on those living nearby is substantial.

In Texas, subsurface mineral rights are legally dominant over surface rights. This means that oil and gas operations, unlike other industrial activities, have an absolute right to intrude on neighborhoods and inhabited areas. Prior to when mineral rights were more widely severed from surface rights, this problem was much less significant.

Development of conventional Texas oilfields occurred largely in an era when surface rights owners willingly accepted the intrusion of drilling because of the financial rewards realized from their mineral rights. Historical development of conventional reservoirs was also primarily in more rural parts of the state, where surface activities impacted far fewer inhabitants.

Methane emissions

One of the real (and potentially harmful) effects of oil and gas operations is from natural gases that can (and do) leak into the atmosphere. Responsibly-conducted operations using established best practices are designed to minimize gas leakage, especially during the production phase of development.

In the not-too-distant past, many crude oil storage tanks were left open at the top, allowing vapors containing large amounts of natural gas to escape into the atmosphere. Operating companies didn't fully comprehend the magnitude of the economic value of their lost product. When environmental regulations ultimately outlawed open tanks, many companies found that the value of the captured gas more than offset the cost of the conversion to covered tanks.

In the early years of oil production, the US had little infrastructure for collecting, distributing, and utilizing natural gas. Production in that era was focused almost solely on oil. But since oil releases a significant amount of dissolved natural gas when it comes to the surface, *associated gas* had to be disposed of in order for the oil to be produced and sold. Flaring (burning of the released gas to the atmosphere) was the solution.

One of the most significant accomplishments of Commissioner William Murray (the **only** petroleum engineer **ever** to serve on the Railroad Commission) was his success in convincing the Commission to outlaw widespread flaring of natural gas.

The struggle to make this happen took many years but finally succeeded in 1949.

Today the Texas Railroad Commission allows the flaring of gas for short periods of time, for example, to test a well's productive capacity. Flaring for extended periods requires special permitting, and is only allowed for such things as productivity testing, equipment maintenance requirements, or insufficient oil pipeline capacity.

Leakage of natural gases into the atmosphere is not something peculiar to shale production or to the hydraulic fracturing process. Some leakage can and does occur during the drilling of all wells. There are methods that can be used to minimize leakage. Increasing the efficacy of these methods unfortunately adds to the time and cost to drill and complete a well. Such methods are thus not always applied.

Once a well is put on production, it ostensibly produces into a closed system. Leakage under such conditions should be minimal. When it is not, invariably it is either due to negligence, or probably more likely, unawareness (gas flowing through production systems is not measured at every point). To the extent that methane concentrations around oil and gas operations are high, it is a problem that can be solved by improved maintenance.

Fracking and water

Concerns about water and fracking come from two sources, both related to the amount of water required to hydraulically fracture a well. The first concern is the large amount of water needed by a fracking operation itself. Though the amount can vary, today something on the order of five million gallons of water are needed per well. And even though this seems like a large amount, total water usage for hydraulic fracturing actually pales against other industrial uses.

Oil and gas, even during the boom years, still consumed only about 1% of Texas' total water usage. Golf courses in Texas required more water than fracking. This is not to diminish the fact that in some drier counties where hydraulic fracturing was very active, fracking operations accounted for a much larger fraction of total water usage.

As Texas comes out of its long drought, and as fracking reduces around the state, water usage issues are likely to subside a bit. However, issues are bound to arise again as shale production operations continue, especially when oil and gas prices return to more commercially-viable levels.

Several solutions have been proposed (and are being pursued) to address the large amounts of water needed for fracking. One of the more interesting proposals is to allow a more widespread free market for fresh water. A freer water market allows farmers, for instance, to sell their water rights to oil and gas drillers. Where such an approach has been used, farmers have used water rights compensation payments to invest in water-saving irrigation equipment. This capital investment long outlives local fracking operations, leading potentially to a permanent reduction in water usage for agriculture.

Another approach under development is research into fracturing fluids that don't require as much fresh water. The large amount of brine and brackish water in the subsurface could then be made available for fracking operations without using the precious fresh water needed for human consumption.

Non-water-based fluids are also being researched for fracking. Hydraulic fracturing of at least some wells can be commercially accomplished with nitrogen or carbon dioxide or even hydrocarbon fluids such as propane. The main impediments to the use of these alternative fluids are their efficiency and cost.

And then there's recycling. All wells, including those that have been hydraulically fractured, naturally produce a certain amount of water. Fracked wells produce back a fraction (typically around half) of the water used to complete the well. But nearly all wells also produce native reservoir water as well. Depending on geological circumstances, the amount of native water can range from very little to quite large.

Post-fracking water production thus includes both injected fracturing fluid as well as natural reservoir waters. Neither is suitable for human consumption nor for surface disposal. Produced water is routinely disposed of by injection into deep formations that already contain objectionable natural subsurface brines.

Water recycling is of interest to oil and gas operators as an economic measure (to allay the cost of purchasing fresh water), but would appear to have positive environmental ramifications as well. A potential downside of recycling, however, is that potentially harmful fluids must be transported and processed on the surface. Some believe that deep earth injection may actually be a safer environmental alternative than recycling.

An interesting conflict has arisen between local water districts and operators who wish to inject wastewater into the subsurface. Some water districts now want to protect brackish subsurface waters from wastewater disposal, arguing that brackish waters (unlike high-salinity brines) could potentially be a future target for desalinization. This argument has some merit, particularly with regard to waters with salinities less than seawater.

Groundwater contamination

The other significant water issue affected by oil and gas drilling is protection of surface waters and groundwaters. Regulations are currently in place to prevent surface spills that might contaminate creeks, rivers, and lakes with drilling or produced fluids. Where contamination has occurred, accidents or negligence were invariably the reason.

Oil and gas wells drilled through groundwater layers also have the potential to cause contamination of the groundwater should produced oil, gas, or water leak into the penetrated interval. To prevent such contamination, surface casing (large diameter pipe) is set across groundwater intervals and cemented in place to provide a seal between the pipe and the surrounding rock. Regulations already exist to ensure that the design and implementation of the casing placement is sufficient to prevent leakage. These regulations were, in fact, strengthened by the Railroad Commission in 2014.

During the time oil and gas wells are on production, surface casing leaks are relatively easy to detect through regular inspections. But when production from a well ceases, there is a danger that the casing in an unmonitored or poorly monitored well might corrode or otherwise break down, leading to unknown and uncontrolled leakage into groundwaters. To mitigate this danger, oil and gas operators are required to properly *abandon* terminal wells by placing cement in the wellbore to prevent future fluid movement.

The US is rapidly approaching nearly 3 million oil and gas wells drilled, over one million of which are still producing. Drilling has been going on for well over a century. The low number of reported occurrences of groundwater contamination is testimony that prevention efforts have been largely (even though not universally) successful. A recent report by the US Environmental Protection Agency concurs with this assessment, as does a similar study from researchers at Yale University.

Earthquakes are due to wastewater wells, not fracking

There are increasing concerns that the application of hydraulic fracturing for oil and gas recovery is causing increased earthquake activity in certain parts of the country (particularly North Texas and parts of Oklahoma). There is essentially no evidence that fracturing operations themselves have induced significant earthquake activity. It has, however, been known for decades that injection of large volumes of wastewater into subsurface formations can (though thankfully rarely does) cause significant increases in seismic activity. These earthquakes are rightfully of serious concern, even though the magnitudes of the earthquakes that are generated have yet to be a serious threat to human life.

Texas has at least 7500 wastewater disposal wells and a total of around 35,000 fluid injection wells. The first category includes wells used solely to dispose of waste fluids. The second category encompasses wells used in secondary recovery and pressure maintenance operations. Clearly an overwhelming number of injection wells in the State operate with no apparent significant associated seismic activity.

The obvious question, of course, is what is to be done when earthquakes do arise. How should the citizenry and local property be protected without completely shutting down important economic activity? This issue will be discussed in more detail in the last two chapters.

Even the roads get worse

One of the unfortunate side effects of oil and gas booms is that local infrastructure tends to be quickly overwhelmed by a rapid increase in activity. Where Texas experienced the most recent infrastructure strains were on its highways. Problems were especially severe in rural counties with small highway budgets and few roads. Some county roadways suddenly became heavily traveled by large numbers of commercial trucks (several hundred truckloads of water and sand are often required to frack a single shale well). Road deterioration happened **very** quickly.

Even though oil and gas activity brings much appreciated economic activity to many local areas, local governments often have a difficult time keeping up with road damage, particularly when there is no way to rapidly access funding needed to carry out repair and construction projects.

CHAPTER THREE
The Texas Railroad Commission

When I started working as a young Assistant Professor in the Petroleum Engineering Department at the University of Texas at Austin, my office assignment was in a suite occupied by the Texas Petroleum Research Committee (TPRC). TPRC was funded by the State of Texas through the Railroad Commission. Its charge was to support small oil and gas related research grants at UT and Texas A&M. My first graduate student was supported by such a grant. As an aside, one of my dearest friends (who remains so to this day) was TPRC's executive assistant. She helped me learn to navigate university life.

During those first few years at UT, I also had the pleasure of meeting Bill Murray. Bill received his MS degree in petroleum engineering from UT in 1937 and subsequently went on to work for the Texas Railroad Commission. Bill eventually became a Commissioner, serving from 1947 to 1963. He passed away in 2004 at the age of 89.

TPRC was vetoed in 1987 by Governor Clements, himself a Texas oilman.

This chapter is intended to provide the reader with background into both the historical role and current responsibilities of the Texas Railroad Commission. This material should prepare the way for a discussion of recommended changes in Texas oil and gas regulations to be discussed in Chapter Four.

Commission politics

The Railroad Commission is part of Texas' plural executive, and likely the most important regulatory agency in the entire US whose heads are directly answerable to voters. Its three Commissioners serve rotating six-year terms, and are thus on the Texas general election ballot every two years. The Commission's authority and jurisdiction is probably best encapsulated on their web site:

> *The Railroad Commission of Texas (Commission) is the state agency with primary regulatory jurisdiction over the oil and natural gas industry, pipeline transporters, natural gas and hazardous liquid pipeline industry, natural gas utilities, the LP-gas industry, and coal and uranium surface mining operations. The Commission exists under provisions of the Texas Constitution and exercises its statutory responsibilities under state and federal laws for regulation and enforcement of the state's energy industries. The Commission also has regulatory and enforcement responsibilities under federal law including the Surface Coal Mining Control and Reclamation Act, Safe Drinking Water Act, Pipeline Safety Acts, Resource Conservation Recovery Act, and Clean Water Act.*

The Railroad Commission has four major operational divisions with the following responsibilities.

Oil & Gas Division

- Prevent waste of the State's natural resources
- Protect correlative rights (subsurface oil and gas minerals) of different interest owners
- Prevent groundwater pollution
- Provide for public safety
- Regulate injection wells (under a federal program)

Oversight & Safety Division

- Provide for safety of intrastate pipelines
- Regulate storage, transportation, and use of LPG, CNG, and LNG

- Regulate city-gate and unincorporated area natural gas utility rates

Surface Mining and Reclamation Division

- Regulate surface mining of coal and uranium

Alternative Fuels Research and Education Division

- Public outreach and education regarding use of natural gas, propane, and other alternative fuels

For the 2016-2017 biennium, the Texas Legislature appropriated $87 million per year and 820 full-time employees for the Commission. Only about 15% of the Commission's budget comes from general revenue funds. The rest comes primarily from dedicated taxes, permitting fees, and federal funds.

The Commission regulated railroads before oil and gas

The Texas Railroad Commission was established in 1891 to regulate railroads and other transportation-related activities. It is the oldest regulatory agency in the State. The federal government took over transportation regulation starting in 1984. The last remnants of the Commission's responsibilities over railroads were transferred to the Texas Department of Transportation in 2005. The Railroad Commission currently has no responsibilities for railroads.

Though the first discovery of oil in Texas at Spindletop occurred in 1901, the Railroad Commission's responsibilities were not expanded to oil and gas until 1917. These responsibilities increased dramatically after the discovery of the East Texas field in 1930. From the 1930s to the 70s, during an extended worldwide oil glut, the Commission used its statutory powers to manage the surplus and support oil prices by limiting oil production from Texas fields. During that time, Texas accounted for over 40% of US crude oil production. When excess

US production capacity disappeared, the Railroad Commission was effectively displaced as a market regulator by OPEC in the 1970s.

A series of regulatory crises ensued after the East Texas field was discovered, ultimately establishing the Commission's responsibilities to prorate oil, eventually for both physical and economic reasons. During the many regulatory battles that ensued in the 1930s and 40s, the Commission's overriding goal was essentially to protect the State's oil and gas resources, independent oil companies, and mineral rights owners from unrestrained competition. The rationale for such regulation was that when there is an overly-abundant supply of a discovered (as opposed to manufactured) commodity, unregulated production brings market chaos and wasteful management of shared natural resources.

The Railroad Commission's most significant oil and gas regulatory activities have primarily been "inside baseball" – preventing waste and protecting mineral rights owners' correlative rights (*i.e.*, the right to their fair share of a resource). The Commission does, however, have responsibility to protect groundwater from oil and gas wells, ensure pipeline safety, regulate certain mining activities, and regulate some gas utility rates.

But things are changing for the Commission. Rapid technological advances were responsible for Texas' most recent oil and gas boom that began in 2005. Hydraulic fracturing technologies have unlocked petroleum-bearing rock formations (shales) previously thought to be non-commercial. Shale production portends very different oil and gas operations than those in the past.

Extraction of oil and gas from shales will undoubtedly look less like a traditional explore-develop-produce paradigm and more like manufacturing. In the US, the locations of our shale resources are already well-established. It will take many thousands (perhaps millions) of wells to fully develop these resources. It's not hard to imagine a future of almost cookie-cutter well construction.

Unregulated production will increasingly become unimportant.

And even though a worldwide oil glut again depressed prices and reduced drilling activity starting in late 2014, Texas' shale resources have not been depleted. Extraction of petroleum resources will undoubtedly continue as adjustments are made to changing market realities.

The giant East Texas field

Because of its seminal role in the formation of many Railroad Commission regulatory policies, a discussion of the interesting and somewhat unique character of the East Texas oil field is appropriate. First is its enormous size. The East Texas field covers 220 square miles across parts of five counties (Gregg, Rusk, Upshur, Smith, and Cherokee). The field originally contained 7 billion barrels of oil and had produced over 5.4 billion barrels by 2014.

Unlike earlier-discovered fields that were controlled by one or a few operators, the East Texas field had hundreds of small companies engaged in drilling. Many landowners even divided their holdings into smaller mineral leases to maximize lease payments. Larger oil companies that were operating in other parts of the State were largely left out of the East Texas boom.

The field is contained by what is called a stratigraphic trap. Such traps are defined by the termination of geologic strata rather than by a more common structural high (anticline). Stratigraphic traps, even today, are difficult for exploration geologists to identify and find. The state of exploration knowledge and technologies in the 1930s was not able to identify the scope of the East Texas field. Consequently, early exploration was conducted by small entrepreneurial companies utilizing less "science" than the major companies. These companies operated in an environment where leases were cheap and drilling was relatively shallow.

Denton has nothing on Kilgore

The City of Kilgore was at the center of the boom. Its downtown at one time had more than a thousand active wells. Wells were drilled in the yards of homes, and derrick legs often touched those on adjoining leases. Independent operators felt compelled to drill as quickly as possible to prevent neighboring producers from draining oil from their leases. Less than a year after discovery, the field was producing over one million barrels of oil per day. A new well was being drilled every hour!

The rule of capture is common law inherited from England, which has guided oil and gas development from its beginning. This principle holds that ownership of oil belongs to whoever produces (captures) it. The rule of capture establishes ownership of natural resources such as game animals, groundwater, and oil and gas. The impossibility of precisely knowing the extent to which underground drainage (or game movement) occurs across ownership boundaries makes the rule of capture a practical legal principle.

When a reservoir is produced too fast

Unfortunately, ultimate recoverable oil from fields such as East Texas could be (and in the early days was) harmed by production that was too rapid. The physical principles governing oil production were just beginning to be understood in the 1930s. It eventually became clear that rapid production would cause excessive amounts of dissolved gas to be released and produced (and at that time flared). Excessive *associated gas* production causes rapid reservoir pressure decline. And when field pressure declines too rapidly, this will reduce the amount of oil that could have ultimately been extracted using more modest withdrawal rates.

The other fly in the ointment with the East Texas field was its very strong water drive (natural water invasion into the reservoir). This recovery mechanism was very controversial and poorly understood in the early producing days of the East Texas field. The strong water drive meant that wells that produced closest to the points of water invasion (deepest depths) were harmed earliest by the invading water. Excessively high production rates cause invading water to "bypass" lower-quality intervals, unnecessarily leaving behind oil that might otherwise be produced. Even modern oilfields, such as Ghawar in Saudi Arabia, have experienced this problem.

The many uncertainties associated with expected future performance of East Texas wells meant that both operators and property owners were motivated to produce oil as rapidly as they could.

The magnitude of the rapid oversupply of oil caused by the East Texas discovery created huge downward price pressures. Oil prices quickly fell from $1.10 per barrel to less than $0.10 per barrel. Producers, as might be expected, responded to lower prices by increasing production even further – sending prices even lower. A similar dynamic occurred with the shale boom that ended in late 2014.

During Texas' early oil and gas development years, the Railroad Commission had no authority to restrict production to meet market demand, having authority only to prohibit physical waste and conserve oil and gas. The Commission at that time was unaware of the connection between production rate and ultimate recovery. But as that connection became better understood, well spacing rules and allowable production rates were ultimately established for the East Texas field. These rules were widely disregarded by many operators.

The beginnings of market-demand proration

In 1931, the Texas legislature, during a special session, passed a bill to limit East Texas production to market demand. However, a federal court ruled that this was an illegal attempt to fix prices and invalidated the law. On August 17 of that year Texas Governor Sterling ordered the Texas National Guard and Texas Rangers into the East Texas field to *shut in* (*i.e.*, cease producing) all of its wells to maintain order. Though the field resumed production on September 5 under a new proration order, lawsuits and smuggling followed.

On February 2, 1932, a federal court ruled that martial law in the East Texas field was illegal. Federal courts continued to strike down subsequent proration orders issued by the Commission. On November 12, 1932, Governor Sterling again called a special session of the legislature, which passed another bill allowing for proration based on market demand. East Texas field production was ultimately capped at a rate of 750,000 barrels per day. The federal government eventually also embraced market-demand proration.

After the field's first decade of existence, many of the problems unique to the East Texas field were resolved. The East Texas field subsequently became an important source of oil for the World War II effort.

From the late 1940s into the 60s, the East Texas field was plagued with a *slant-hole* scandal. Some unscrupulous operators had drilled wells from unproductive acreage to locations underneath productive leases. A total of 380 deviated wells were discovered and shut down – after an estimated $100 million worth of oil was stolen. None of the thieves, many leading East Texas citizens, were convicted. Stricter rules governing slant-hole drilling were subsequently adopted by the Railroad Commission.

Preventing waste and protecting correlative rights

The Railroad Commission ceased regulating the State's oil production to meet market demand in 1972 (though it retains the statutory authority to do so). The Commission does, however, continue to prevent waste and protect correlative rights by regulating well spacing and well density, as well as field and individual well *allowable* production rates.

The term *correlative rights* refers to a legal doctrine limiting the rights of owners of a common subsurface resource (e.g., groundwater or oil and gas) to all parties' reasonable share. Correlative rights rules and regulations essentially modify the rule of capture in order to provide mineral owners access to their "fair share".

Well spacing and density rules

Spacing rules are established by the Commission in each field to set the minimum distance between wells and to the boundaries of adjacent tracts of land. Statewide Rule 37 prohibits wells from being closer than 467 feet (approximately the length of 1-1/2 football fields) from property lines and from within 1200 feet (4 football fields) of other wells producing from the same zone. Rule 37 can be modified by seeking special field rules from the Commission. Field rules are subject to technical hearings that can include objections from interested parties.

Density rules are in place to prevent clustering of wells and overproduction. *Density* refers to the minimum amount of acreage that can be assigned to a single well. Statewide Rule 38 provides for a standard density of 40 acres for oil wells and 160 acres for gas wells. This too may be modified by applying for special field rules. Vertical wells drilled on a uniform 40-acre square pattern are 1320 feet (1/4 mile) apart. Wells on a 160-acre square pattern are 2640 feet (1/2 mile) apart.

It has been a challenge for the Railroad Commission to extend regulatory concepts for vertical wells – developed over many decades – to horizontal wells. This is particularly a problem for shale wells in ultra-low-permeability zones that have been massively hydraulically fractured.

Additional acreage may be assigned to horizontal wells under Statewide Rule 86. Standard rules require the same lease-line spacing (467 feet) and between-well spacing (1200 feet). However, the Commission has a formulaic rule that assigns additional acreage in addition to that assigned for vertical wells.

Most horizontally-developed fields in Texas are operated under special field rules, typically having a 330-foot (just over a football field) lease line spacing and no between-well spacing requirements.

The rules for horizontal wells are based on the portion of the well that penetrates the productive horizon, not the surface location. It is not uncommon for shale well operators to completely develop one to two square miles (640 to 1280 acres) with wells on a single pad, typically smaller than five acres or so in size. Off-lease surface locations are also common for acreage developed with horizontal wells.

Pooling

Sometimes mineral owners will voluntarily elect to form pools to develop a set of contiguous properties. A *pool* (or unit) is created by combining separately-owned mineral interests underlying different tracts of land into one entity. Production proceeds are allocated among the various owners on an agreed-to proportion based on acreage, well productivity, estimated recoverable oil, etc. The operator assigned to produce pooled acreage is relieved of spacing requirements associated with lease boundaries. Some states, but not Texas, have provisions for mandatory pooling.

Allowables

Two forms of allowed production rates (*allowables*) are regulated by the Railroad Commission. First, an allowable for a field is determined using a concept called *Maximum Efficient Rate (MER)*. Statewide Rule 45 provides for the calculation of a "yardstick" allowable based upon the depth of the reservoir and the proration unit size. As with other field rules, operators can apply for modification where they believe the MER should be greater than that determined by Rule 45.

Once a field allowable has been determined, individual well allowables are prorated to protect individual mineral owners' correlative rights. Well allowables are based on each well's productive capacity, assigned acreage, reservoir mechanics, and past production.

The Railroad Commission requires operators to report oil, gas, and water production volumes on a regular basis in order to monitor operators' adherence to field and well allowables.

Injection and disposal wells

The Texas Railroad Commission also regulates injection and wastewater disposal wells under a federally-approved program. Fluids may be injected into productive reservoirs for enhanced recovery projects to increase production. In such intervals, both fluid injection and extraction is done simultaneously.

Other injection wells are permitted to dispose of wastewater from oil and gas operations into deep non-productive intervals. These wells are the ones primarily responsible for earthquake activity that has been experienced in the State.

Groundwater protection

Protection of groundwaters falls under the Railroad Commission's purview in two ways. First, it is responsible for mitigating and preventing pollution from fluids handled on the surface, including those used during both drilling and production operations. The Commission also protects groundwater through regulations designed to ensure that all wells have sufficiently well-designed and constructed surface casings to prevent leakage from deeper productive intervals into groundwater intervals.

In 2014 the Commission strengthened its Rule 13 requirements to better protect groundwater where the risks may be higher than in normal production situations, including wells in which hydraulic fracturing treatments are to be applied. Those rules are highly technical in nature, and are claimed by the Commission to reflect, in their words:

> *... many best management practices that are already being implemented by most operators.*

The Commission also operates an abandoned well plugging and site remediation program. When wells cease being productive, there are requirements to properly plug and abandon them in order to prevent future contamination by leakage from deeper zones. Defunct operators sometimes leave behind *orphaned wells* that have not been plugged, and are consequently a risk for future groundwater contamination.

The Commission's orphan well program uses funds provided by the oil and gas industry through fees and taxes. Wells and sites are remediated with these funds when responsible operators cannot be located.

Pipeline safety

Texas has over 400,000 miles of pipeline representing about a sixth of total US pipeline mileage. Pipelines in the State carry intrastate oil and gas production, liquefied petroleum gases, and hazardous fluids, and are used as field gathering lines. The Railroad Commission has safety responsibility over many, though not all, of these categories.

The Railroad Commission's Pipeline Safety department enforces federal and state laws and regulations for pipelines. The Commission also enforces an underground pipeline damage prevention program designed to prevent damage from excavation around pipeline facilities.

Alternative fuels (but not alternative energy)

The Railroad Commission operates programs related to liquefied petroleum gas (LPG), compressed natural gas (CNG) and liquefied natural gas (LNG). In addition to enforcing statutes and regulations designed to ensure safety, an Alternative Fuels Research and Education Division operates education, outreach, and incentive programs related to alternative fuels.

Natural gas rate setting

The Railroad Commission has jurisdiction over natural gas utility rates in certain areas outside of municipalities. The Commission also has jurisdiction over rates for utilities that deliver gas to a distribution utility (*city-gate*). Jurisdiction over other natural gas utility rates falls under the Texas Public Utility Commission.

Coal and uranium mining

In 2013, Texas was the largest consumer of coal, as well as the sixth largest coal-producing state. The Railroad Commission has jurisdiction to regulate surface mining for both coal and uranium.

Though there is no surface mining for uranium currently underway in the State, exploration activities related to *in-situ* uranium mining is overseen as well. The Commission also administers a program to reclaim abandoned mine sites that were established prior to a 1975 federal surface mining law.

Challenges ahead

Despite its misleading name, Texas Railroad Commission is charged with regulating operations that produce the State's mineral resources, primarily oil and gas but extending to mined-resources as well. As the shale boom continues in the future, the Commission's role in regulating oil and gas will continue to be of significant importance to the State.

The next and final chapter of this book is devoted to a consideration of regulation. The focus of the chapter will be on oil and gas and the Texas Railroad Commission. The reader will hopefully find that there are also lessons from this discussion that could be applied to other regulatory arenas as well.

CHAPTER FOUR
How to Regulate a Free Society

During its 84th Session, the Texas Legislature passed, and the Governor signed, House Bill 40 in response to the fracking ban approved by City of Denton voters in a November 2014 referendum vote. The Denton referendum was a local voter-led initiative. HB 40 was then, and remains still, highly controversial. The bill puts limitations on the ability of local governments to regulate oil and gas operations within their jurisdictions. HB 40 effectively overturned the Denton vote.

Extensive testimony was given at the House Energy Resources Committee public hearing for HB 40. I happened to sit next to Grand Prairie's City Attorney while we waited for the hearing to begin. The discussion that ensued as we waited revolved around the fact that local fracking issues were really about a conflict of rights. Mineral owners have rights to realize the economic value of their property, while surface owners have rights to the quiet enjoyment of their property. The conflict arises when one essentially nullifies the other.

Grand Prairie's City Attorney was there to testify against HB 40. Grand Prairie had already wrestled with the conflicting rights issue. While admitting that negotiations had been difficult, long, and painful, he believed that they had been successful. Grand Prairie's regulations involved a combination of *setbacks* (minimum allowed distance between well sites and homes), restricted road access (both time scheduling and specific roadways), and road-damage bonds (the city even took before-and-after photos to document road damage).

Grand Prairie's City Attorney noted that his city was highly industrialized. It was clear to him that what worked in Grand Prairie would **not** necessarily work well for other communities. His strong belief was that *local conditions should guide local regulations.*

The Texas Railroad Commission is a great place to start

This book would not be complete without recommendations for ways forward. This last chapter will provide reflections on, and recommendations for, regulating oil and gas in Texas. There is no agency more relevant to a discussion about regulatory policy than the Texas Railroad Commission:

- **The stakes are high.** Activities regulated by the Railroad Commission generate huge revenues for Texas citizens as well as the State. At the same time, opportunities for serious personal and environmental damage are also high.
- **Many regulatory issues involve conflicting rights.** The three primary private actors affected by oil and gas operations are surface owners, mineral owners, and operating companies. All have rights as well as responsibilities that must be considered.
- **Texas is an oil and gas producing province of worldwide significance.** Texas has dominated historical US production, accounting even today for a quarter to a third of all oil and gas produced in the US. While extraction of hydrocarbons from shales has yet to expand to the rest of the world, Texas is leading the way in the US.
- **The Texas Railroad Commission is directly answerable to voters.** The Railroad Commission is fairly unique among regulatory agencies in the US, particularly with regard to its politics. How the Commission will respond to its regulatory challenges has yet to be fully realized.

What is means to regulate

A discussion of regulatory policy should begin by first considering why regulations are needed in the first place. And if needed, what form should those regulations take? As will become clear from the ensuing discussion, the issue to be addressed is how personal protections can be ensured by law and regulation in a way that does not deprive us of the very rights, freedoms, and property that government is charged with protecting.

One kind of regulation, of course, is legislation. There are mechanisms, though imperfect, that hold originators of legislation accountable. When details are too complicated to be effectively considered by legislatures, broad-brush rules are often passed that prescribe regulatory agencies such as the Railroad Commission to write and administer regulations. This essentially moves regulated activity from the legislative branch's jurisdiction to the executive branch.

Regulations have the force of law, but are administered by designated regulators. Regulations can obviously be overly complex and prevent innovation, but they usually are at least clear. Rule-making and mechanisms for fixing regulations, as frustrating as they might be, are more accessible than legislative action. And as a final resort, decisions by regulators can be appealed to legislatures and/or the judicial system.

When legislators grant broad discretionary powers to specific regulators, there is a danger of inadequate checks and balances that normally serve to prevent executive branch abuse of power. This is especially the case when vague responsibilities are granted with no clear objective standards. Though this is not the case with the Railroad Commission, most regulators are only indirectly answerable to voters by reporting through the head of the executive branch. Oftentimes, regulatory agencies essentially act as a "state within a state".

At the risk of oversimplification, let's begin the discussion by assuming that the primary purpose of regulation (including legislation) is either to cause good outcomes or to discourage bad ones. Let's further simplify the discussion by dividing regulations into those that are based on a command-and-control (CAC) paradigm vs. those that are based on a retribution-and-restitution (RAR) paradigm.

CAC regulations are those that seek to ensure outcomes through such things as code requirements, prior approvals, and inspections. In the absence of fraud, CAC regulations achieve compliance through forced behaviors before an action is allowed. RAR regulations, on the other hand, are those that seek to ensure outcomes through after-the-fact remedies such as punishment or tort actions. RAR regulations achieve compliance by establishing legal standards for harm, for which remedies can be imposed if violated.

Many believe, as I do, that governments (including both legislators and regulators) should restrict their powers to only those necessary to protect our liberties and our property. These protections should encompass personal liberties, public safety, access to our commonly-shared air and water resources, property rights, and our right to engage in free and prosperous commerce. It is unfortunate that we sometimes also need to be protected from the power of governments as well as from each other.

Giant governments as well as private giants are both particularly good at impinging on our rights and our property. We are in the most danger when these two giants collude. And watch out when the taking of individual liberties or property is justified in the name of a broadly dispersed and vaguely perceived common good.

Some who read this book may wish to imagine some future Utopian state where freedom and liberty are ideally maximized for all. That vision won't be provided here. Rather, a more pragmatic approach is presented that focuses on a few definite, achievable, and practical actions that could be taken by the Railroad Commission and the State of Texas to move us toward a freer future. I hope that readers might be convinced that the recommendations contained in this chapter are worth considering and perhaps pursuing, even if falling short of perfection.

Downsides of command-and-control

The explosion of command-and-control regulations in the US in recent years may in large part be attributable to (sometimes unspoken) collusions between private and public giants. It is difficult for those "in-charge" to resist the temptation for command-and-control. Though Republicans regularly extol the virtues of limited government, they still want intrusive inspections before we board airplanes and the ability to access our private phone conversations. Democrats, while believing that we should marry whom we wish and that women should have abortion rights, also seem to prefer taking away our gun rights, restricting our healthcare choices, and even limiting the food we wish to consume.

Large private interests (*e.g.*, corporations) often prefer the certainty of federally-mandated command-and-control regulations to the uncertainty of state and local regulations, or (heaven forbid) exposure to individual restitution torts. Though commercial interests often regale against government "intrusion", they are also too often more than willing to embrace financial incentives (a weak form of CAC) for their preferred practices.

CAC regulations would seem to be the least desirable of the two regulatory approaches. Such regulations inhibit innovative methods (even those that are less costly and more effective) for solving regulatory objectives. The nature of both laws and regulations is that they are slow to change, and must be considered by legislators and bureaucrats who don't have always have the requisite knowledge or expertise to oversee either their suitability or their proper implementation. The natural bureaucratic inclination for "more is better" tends to stimulate ever-increasing levels of governmentally-mandated command-and-control. It is extremely difficult to overcome the illusion of **actual** control. And what (tongue-in-cheek) could possibly go wrong with commanding that the "right thing" be done? Lessons from history abound.

Another significant problem with CAC regulations is that they enable bad actors to claim they "followed the rules" as an excuse for bad outcomes. It is far too easy for both judges and juries to presume "they didn't do anything wrong" when CAC regulations are properly followed. It is a natural human assumption that when someone does what they are told by the government that they should not be culpable.

The Denbury Pipeline case (discussed in a later section on Common Carrier Pipelines) is a good example of this problem. Denbury Green Pipeline Company was sued for eminent domain abuse facilitated by their assertion of common carrier pipeline status. Denbury successfully argued in two lower courts that common-carrier status was effectively **approved** when it received a pipeline permit from the Texas Railroad Commission that included a check-box so asserting. The Texas Supreme Court fortunately overruled the lower courts in this case, stating that Texas law provided the requirements for common-carrier status, not a check-box on a Railroad Commission permit!

Arguments for continuing CAC regulations are usually based on their efficacy, even though this efficacy is more perceived than real. A more legitimate fundamental question, though, is whether there are situations where undesirable outcomes are so severe that there is simply too much risk to take by **not** commanding. "Too-risky" might include outcomes that result in death, or that would result in the bankruptcy of an individual or company such that restitution would not be possible in the future. It will be important to keep this issue before us as we explore regulatory alternatives for oil and gas.

Retribution and restitution would be better

As the reader might already guess, my bias is toward using RAR regulations where at all possible. Anticipation of restitution for bad outcomes will cause rational behaviors by individuals and companies that act to effectively mitigate possible negative consequences. And where rational behavior fails to materialize, remedies based on restitution provide compensation for those negatively impacted by harmful behaviors.

Some might argue that adding significantly more RAR approaches to regulation would simply clog up the courts and enrich trial lawyers. With our current judicial and regulatory regimes, this could well turn out to be the case. Should the State of Texas embark on a system more oriented to RAR regulations, creative ideas for change will certainly be needed. A few of these ideas are discussed in the following sections. Some restraint of the legal profession might even be in order.

After the British Petroleum Gulf of Mexico oil spill disaster in 2010, restitution occurred through disaster funds administered by an independent third party. By all accounts, the method of administering these funds had positive outcomes, quickly getting money to those most affected by the disasters without over-burdening a slow-moving judicial system. Though some funds were poorly distributed to undeserving recipients, this is still a model that deserves further consideration. At the very least, BP's financial reparations have already engendered increased diligence by other offshore oil and gas operators.

Start with a cultural shift

It is important to recognize that both CAC and RAR regulatory approaches will be imperfectly applied. Undoubtedly arguments will be made that **more** of the failed approaches of the past will work better. Unfortunately history tells us that more government is unlikely to be better government. We should recognize the **decreasing** ability of government to effectively command and control an **increasingly** diverse and complex society. We should recognize that individual action collectively exercised is most likely to create the best outcomes. We should recognize that a society that values freedom and personal responsibility is most likely to be the one we wish to live in. When deciding on how to pursue regulatory policy, we should be certain that we maintain government's focus on its primary **protective** responsibilities. We should resist the temptation to command and control society into some preordained vision of someone's idea of perfection.

Three fundamental cultural shifts would seem to provide the most benefit to our regulatory regimes – transparency, regulatory reform, and a focus on personal liberty.

Transparency is essential for any robust democratic society. To engage in voting and in public discourse, citizens must be able to judge for themselves whether their agents and representatives are indeed working in their best interests. Change cannot and will not occur without transparency. Regulatory agencies, in particular, should adhere to strict transparency standards. Regulatory agencies are charged with carrying out the people's work. The people, not just their legislators, deserve to have a full and open view of what is being done. Voters should demand transparency.

Regulatory reform is another essential activity. Governmental intrusion into our commercial and personal lives has, by now, a long history. It should be no surprise that many regulations and laws are outdated, inefficient, and badly in need of overhaul. A continuous focus on reform and change is the only way that regulations can be refreshed, revitalized, and minimized.

Finally, regulatory policy must be based fundamentally on the preservation of liberty. With every new regulation, every re-written regulation, every consideration to do away with a regulation, the question must always be asked as to how doing so will alter personal liberties and property rights.

We must fight for a regulatory culture that resists the dubious contention that a small amount of good spread across a large number of people is preferable to a large amount of harm to a small number of people. This is a morally bankrupt position that is contrary to the American and Texan traditions of protecting minority and personal rights.

The remainder of this chapter provides examples of changes for Texas to consider in its regulation of oil and gas operations based on these principles. Though the following discussion focuses on oil and gas, there are clearly implications for other areas of endeavor as well.

Transparency

Every regulatory agency is prone to *regulatory capture.* This term refers to a form of political corruption that occurs when an agency becomes overly concerned with the special interests of groups it regulates at the expense of the general public. One might expect an agency with elected heads to be less prone to regulatory capture, especially one for which voters have the opportunity to make a change every two years. Alas, this is not the case.

The Texas Railroad Commission is a case in point.

Texas voters are amazingly uninformed about the importance of their Railroad Commission. Ryan Sitton, the Commissioner elected in 2014, stated that his campaign polling revealed that *less than 5% of Texans were aware that the Railroad Commission regulated oil and gas.* Even when voters are made aware of the Commission, they remain confused about why it continues to have such a misleading name. Voters that might otherwise be concerned with changes in how the State of Texas regulates oil and gas are left unaware of their potential to improve things.

There have been periodic conversations about the Railroad Commission's name and its "historical" significance. Some even worry about whether powers delegated to the Commission by the federal government would survive a name change. Sadly, others seem to simply be seeking to obscure the nature of the Commission through its misleading name.

It is historically understandable how the Texas Railroad Commission arrived at this level of obscurity. Authority over oil and gas operations has been gradually and incrementally added to the Commission's responsibilities over an extended period of time. Once the Commission had authority over oil and gas, giving it additional duties was both convenient and efficient.

Add to this the fact that most of the Commission's historical duties have been "inside baseball", effectively settling fundamental conflicts between mineral rights owners, independent oil and gas companies, major companies, and even pipeline operators. Is it any wonder that the players with the most at stake seek to influence those who run this important agency? The broader public stake in the Commission is sometimes far less obvious.

Railroad Commissioners openly admit to their dual role as oil and gas industry champions as well as the industry's chief regulator. This dual role is effectively an open admission of enshrined regulatory capture, and an open admission of government-business cronyism. This dual role as regulator and industry champion causes, as it should, a large number of voters to worry about Commissioners being too cozy with the industry they regulate. In this environment, the primary qualifications for getting elected Railroad Commissioner are name recognition and party affiliation. Consequently, there are essentially no significant policy debates held during election years.

But things are changing. The shale revolution enabled by hydraulic fracturing created a new oil and gas boom in Texas. Broad public interest in oil and gas matters is unlikely to remain as low as it has in the past. Public fears about fracking combined with the increasing scope of oil and gas operations due to the ever-expanding geographic footprint of shale production, promise a future when more and more Texas voters will be affected and will demand something different from their Commission.

It is imperative that the Railroad Commission move past its image as primarily serving the oil and gas industry, to one that seeks to serve all Texans.

Ensuring public trust

The Texas Sunset Advisory Commission is a good place to start as we seek to ensure public trust. Texas statute requires that every State agency undergo a sunset review every ten years. Each agency must be re-authorized on a ten-year rotating schedule. The Texas Sunset Advisory Commission is composed of a body of lawmakers that prepares re-authorization recommendations for the Legislature to consider.

As of 2015, the Sunset Commission is about to begin work on its third report on the Railroad Commission. Both the 83rd (2013) and 84th (2015) Texas Legislatures failed to re-authorize the Commission on its regular ten-year cycle. The Commission's reauthorization will thus again be considered by the 85th Legislature.

Past Sunset Commission reports have consistently recommended a number of significant changes to the Commission to enhance its transparency with Texas voters. Included in these recommendations were:

- Changes to the Commission's name to better reflect its actual role as an oil and gas regulatory agency
- Limitations on campaign fund-raising by sitting Commissioners
- A robust recusal policy to ensure that conflicts of interest, real or perceived, are eliminated

Regulatory agencies have a special responsibility to be seen as fair arbiters and enforcers of regulations and policies that have been enacted through the legislative process. Even the industry being regulated certainly must understand how important it is that Texans believe that their collective interests are being properly stewarded.

The Railroad Commission is far too important an agency to remain obscure or be mistrusted. Given the impact its decisions have on important economic and environmental matters, maximum transparency of its operations would best serve the people of Texas. Following Sunset Commission recommendations would be a great place to start.

Regulatory reform

Regulations that are overly complex, outdated, and obscure typically favor large well-connected entities over smaller entities and the general public. Regulations that have been on the books for decades are obvious targets to be modernized, clarified, and in some cases eliminated. Texas Railroad Commission regulations are certainly in such a state. The Commission has been promulgating oil and gas regulations for a century now. A good house cleaning is certainly in order.

Regulatory reform should start by establishing a culture of review and regular procedures for ensuring continuous re-evaluation. A sunset review policy could target something like 10% of all regulations per year, starting with the oldest ones first. Such a policy would not preclude changing regulations when serious needs arise. It would simply require that **all** regulations undergo a regular process of re-consideration.

What exactly should reconsideration entail? What should be the criteria for changing or eliminating regulations that are already on the books? Here are a few guidelines to consider.

- **Every regulation should have a clear objective in terms of public good.** That objective should be clearly stated in publicly available documentation.
- **The rationale for efficacy behind every regulation should be clearly and publicly stated.** Merely requiring some action without such an explanation is not sufficient. Clear and simple explanations of what each regulation is expected to accomplish should replace bureaucratic and legalistic verbiage.
- **Dispersed public goods should be properly weighed against concentrated adverse effects on regulated entities.** It is far too easy to make "greater good" arguments without considering adverse effects on individual liberties, property, and free commerce.
- **Where feasible, industry standards should substitute for written regulations.** Building codes fall into this category. Industry associations such as the American Petroleum Institute already promulgate best practices for their industry. If industry

best practices are insufficient, documentation as to where they are deficient provides both a rationale for regulations as well as input for industry associations to consider.

- **Retribution-and-restitution regulations should be favored over command-and-control.** Wherever possible, regulated entities should have the flexibility and authority to find ways to prevent bad outcomes, while still assuring liability for bad outcomes or behaviors.

Devolution

A free society is one where individuals, not the government or its agents, make the most important decisions. Where collective actions are necessary, voluntary transactions and agreements between individuals and groups of individuals ensure the most mutually-beneficial outcomes.

Some would argue that a more perfect world could be constructed around some form of anarchistic volunteerism. However, this is not the world in which we currently live. For better or worse, our nation and our states have been organized through a complex system of constitutional requirements, legislated statutes, regulations, and various levels of governmental authority. It makes the most practical sense to focus on how we move from our current condition, rather than focusing too much on an ideal future one.

Spontaneous order among humans naturally springs from complex and voluntary transactional networks. Humans are social animals whose brains have evolved to support highly-complex social interactions. Those interactions were more easily managed when we lived in smaller tribal groups. There are even arguments put forth by anthropologists that our brains can only handle persistent social relationships in groups that are roughly the size of a military company (100-250 people). This concept is known as Dunbar's number.

As civilization has evolved, it has required more and more complex organizations that outgrew tribes, culminating in the nation states that make up today's world. The success of nation states as organizational structures has led some to believe that more and more top-down control is needed to deal with the increasing complexity of the world we live in. That is unlikely to be true.

The complex world we live in already requires a great deal of widely-dispersed transactional trust. Our employer transfers money to our bank that we withdraw at will. Excellent food choices show up at our grocery stores, giving us a wide choice of variety, price, and quality. The energy we need for our homes and our transportation is there at the flip of a switch or activation of a gasoline pump.

Arthur C. Clarke prophetically told us that: "Any sufficiently advanced technology is indiscernible from magic." If one steps back and wonders at the complexity of the advanced technological world we live in, it is nothing short of magical. *This magic is too important to be left in the hands of a few magicians.*

One of the most complex and well-functioning organizations that has been studied by organizational theorists is that employed on US naval aircraft carriers. The characteristics of aircraft carrier organizations are well documented in an interesting study published in 1987 entitled: "The Self-Designing High-Reliability Organization: Aircraft Carrier Flight Operations at Sea".

On the face of it, organizing a modern aircraft carrier has many things going against it – thousands of personnel carrying out mission-critical tasks, many of which could result in death if done incorrectly. Add to that the high level of turnover and the diversity of tasks required on a carrier, and one might conclude it was an impossible organization to establish and manage. But it's not. This organization is remarkably successful for a few key reasons.

- **Dispersed authority and responsibility.** Though there is a formal top-down chain of command, individual officers and sailors have a high degree of authority over their individual areas of responsibility. From the above study:

  Even the lowest rating on the deck has not only the authority but the obligation to suspend flight operations immediately, under the proper circumstances, without first clearing it with superiors. Although his judgment may later be reviewed or even criticized, he will not be penalized for being wrong and will often be publicly congratulated if he is right.

- **High turnover.** Paradoxically, what would appear to be a negative turns out to be a positive. Moving personnel from position to position actually allows a great deal of flexibility, especially under adverse conditions. Naval personnel are purposefully assigned to multiple job classifications, usually on multiple ship assignments. This creates the ability of the organization to respond to crises or suddenly unavailable specialties with a minimum of disruption and a maximum of efficiency.

- **Redundancy.** Redundancy allows for critical units and components to continue to function in ways that create a high-reliability organization. Although physical redundancy is important, so is the ability of an organization to adapt under stress. Most of the personnel on carriers are familiar with several tasks and have the ability to execute them in an emergency.

This is a fantastic model for how complex societies can, and actually do, operate. One of the beauties of our federalist system's bias toward the lowest levels of government and individual authority is that this model is already built in. Unfortunately, the increasing power of nation-states has meant that we too often look to top-down and command-and-control regulatory structures to solve our problems.

As we do so, these structures actually erode our society's ability to adequately deal with the complex mission-critical issues not too dissimilar from those faced on aircraft carriers. The temptation of top-down structures is their supposed efficiency. Redundancy seemingly means wasted resources. Turnover seemingly means poorly-trained workers. And dispersed authority seemingly means lack of control. Yet aircraft carriers found just the opposite!

At the risk of using a potentially negative word, I'm going to use the word *devolution* to signify steps to enable more decisions to be made by individuals, private organizations, or the lowest levels of government. Devolution usually means moving powers from more centralized to less centralized governing authorities. In this section, however, I will also discuss devolution in terms of moving powers away from government and to personal and private organizational action.

The problem with the word, of course, is its somewhat negative connotation as "backward evolution". Those who believe that evolution inexorably moves things from being primitive to being advanced will fear that devolution is a step backward. Perhaps in some cases this is true. But sometimes steps backward provide the way forward.

In nature, plants and animals that are overly specialized to a particular environment will more than likely become extinct when that environment changes. Top-down approaches to organizations almost invariably lead to rigid and sclerotic ways of governing. Governments (and other large organizations) end up being poor at adapting to new situations.

Just as in nature, when there are multiple organizations and multiple individuals engaged, the sclerotic failures of a few don't result in extinction. When government retains its monopoly on action, rigid and sclerotic becomes the only game in town. Those that desire a strong central state often fail to take into consideration such failures from the past.

It could even be argued that today's levels of centralization have actually led to more chaos, not less. Massive increases in laws (many that aren't enforced) create an ever-expanding legal terrain that is increasingly harmful to individual action and innovation. Devolution is the way out.

Devolution in practice

And even those that ostensibly favor free and diverse markets often prefer strictures that are in fact less free and less diverse. Some of our largest and most powerful companies often prefer a strong central government to have more power over their actions, rather than either local governments or individuals. You have heard the arguments:

- "We need to have uniform and predictable regulations across the State."
- "The State's broad economic benefit is more important than local concerns."
- "Things just work better this way."

Often what they really mean is that it's much easier to lobby one governmental body than hundreds.

To adequately deal with the complexities of our modern world, we need to let go of our illusion of control. And then let go of our fear of individual and local power. Here are a few suggestions for doing so.

- **Diminish the power imbalances that currently exist between large commercial entities and individuals.** For example, the dominance of mineral rights over surface rights, along with eminent domain authority given to common carrier pipelines, create power imbalances that all too often leave individual land and home owners powerless to protect their property.

- **Where there are principally local effects, allow principally local control.** This issue will be addressed in more detail in the section below on local fracking bans. There should always be **very** compelling reasons for higher levels of government to preempt local control. Such reasons would certainly include protection of personal liberties and property rights. Anything beyond those reasons should be considered with great suspicion.

- **When in doubt, move authority downward.** Where some governmental action is deemed necessary, action at local levels should be preferred. The more local the government, the more responsive it is to voter action. In Texas, local initiatives and referendums are key ways for voters to influence their local governments. These procedures are not available at the state level.

It goes without saying that devolution should not apply to our fundamental liberties and property rights. The *US Constitution's Bill of Rights* enumerates many basic rights that neither lower levels of government nor individuals should be allowed to diminish. Even though we consider certain rights to be inviolate, exactly what those basic rights are should continue to be the subject of intense political debate.

Recommendations for Texas

The following sections describe specific recommendations for improving oil and gas related regulations in Texas. These issues range from earthquakes to local fracking bans. Though the focus of this book has been on the Texas Railroad Commission, many of the recommendations would require legislative or perhaps even local governmental action.

The particular issues discussed here should be considered as examples of how regulatory regimes can be established that are effective protectors of individual liberties, property rights, and our common natural resources. Though some of the recommendations might entail a significant amount of political "stretch", they are designed to have sufficient pragmatic possibilities to engender further consideration and discussion. Hopefully these proposals will provide the basis for additional public dialog on oil and gas regulations in Texas. Perhaps the reader will also be able to imagine how the ideas explored here could be extended to other regulatory areas as well.

Human-induced earthquakes

Concerns are on the rise that the application of hydraulic fracturing for oil and gas recovery is causing increased earthquake activity in certain parts of the country, including Texas. There is scant evidence that fracturing operations themselves induce significant earthquake activity. There is, however, correlative (though not irrefutable direct causative) evidence that injection of large volumes of wastewater into subsurface formations can, in some relatively rare circumstances, cause significant increases in seismic activity. Even though these earthquakes are rightfully of serious concern, the magnitudes of the earthquakes that are generated have yet to be a threat to human life.

Texas has at least 7500 waste disposal wells and a total of around 35,000 fluid injection wells. The former category includes wells used solely to dispose of waste fluids, while the latter category encompasses wells used in secondary recovery and pressure maintenance operations. Overwhelming numbers of waste injection wells operate across the State with no apparent significant seismic activity.

The obvious question, of course, is what should be done about the problems such as those that occurred near Azle, Texas, in 2013. How can local property rights be protected without completely shutting down important economic activity? There are both near-term and long-term questions that should be addressed with regard to the regulation of wastewater disposal wells.

Making a strong scientific case for a causative link between a particular human activity and a particular sequence of earthquakes typically involves special studies devoted specifically to the question. The first step, of course, is to ascertain that there is ample evidence to support the conclusion that wastewater disposal could be the cause of the Azle earthquakes.

A study published in 2015 by researchers at Southern Methodist University (SMU), the University of Texas at Austin, and the US Geological Survey (USGS) concluded that earthquake activity in late 2013 around Azle was most likely related to injection into two particular wastewater injection wells. The study concluded that:

> *... pore-pressure models demonstrate that a combination of brine production and wastewater injection near the fault generated subsurface pressures sufficient to induce earthquakes on near-critically stressed faults. On the basis of modelling results and the absence of historical earthquakes near Azle, brine production combined with wastewater disposal represent the most likely cause of recent seismicity near Azle.*
>
> *Critics of these [prior] studies note, correctly, that tens of thousands of currently active injection wells apparently do not induce earthquakes or at least not earthquakes large enough to be felt or recorded by seismic networks. Why some injection wells induce seismicity while others do not is unclear.*

The authors of the SMU study present a credible case for concluding that wastewater injection can produce conditions that will induce at least minor earthquakes. They also make a credible case that subsurface activity was the most likely cause of the earthquake activity near Azle.

In my opinion, the correlative link between wastewater injection and seismic activity was fairly clear from the performance history of the well nearest the fault where activity occurred. Seismic activity began shortly after the injection rate into that well was increased by approximately 50%. Activity then ceased when the injection rate returned to its pre-earthquake levels in late December 2014. No detectable earthquake activity was experienced when injection into the same well was previously at even higher levels. These observations lead to the following questions that need to be answered.

- Seismic activity seems to clearly have been triggered by some combination of time (cumulative injected volume), injection rate, and pressure. It is important to better understand what combinations of those factors could have induced earthquakes.
- Given the large number of wastewater injection wells in Texas and elsewhere, there still remains a question as to what specific subsurface geologic conditions make some areas more prone to induced seismicity than others.

Topics such as the connection between earthquakes and water injection require regulatory oversight that is capable of understanding how complex technological and economic issues should be resolved. With the growing need for wastewater disposal in Texas, it is imperative that the Railroad Commission give this issue the serious attention that it deserves. The actions taken by the Commission in 2014, unfortunately, were insufficient.

In early 2014, the Commission hired a seismologist. Though having additional expertise at the Commission might seem wise, it is hard to understand why the Commission could not have simply reached out to Texas' highly experienced and competent geological and geophysical scientific community for help (with much lower cost and much higher credibility). The premier expert on Texas earthquakes works just up the road at the University of Texas Institute for Geophysics!

In late 2014, the Commission promulgated additional regulations that were intended to reduce the risk of human-induced earthquake activity around the State. Unfortunately, the lack of scientific and engineering rigor inherent in these changes makes them appear as simply temporary stop-gap measures, intended more to assuage public fear than to solve any real problems.

Five key aspects were included in the new regulations: 1) identification of a *zone of influence* around injection wells, 2) reporting of historical seismic activity within the zone of influence, 3) specifying that additional data may be required from well operators should the Commission determine there is earthquake risk, 4) provisions for additional data reporting that may be required by the Commission, and 5) provisions that a water disposal permit may be modified, suspended or terminated should a link with seismic activity be either suspected or shown.

The problem with the new regulations is that there was absolutely no discussion as to whether the earthquakes near Azle could have been prevented had these regulations been in place in 2013. Without such evidence, it would appear that regulations have, once again, been put in place that provides an appearance of solution in lieu of an actual one.

In 2015, the Railroad Commission investigated earthquakes that had occurred in Johnson County. The Commission required nearby wastewater well operators to submit additional data, including "falloff tests", to determine the likelihood of injection-induced earthquakes. The following statement was issued by the Commission upon completing its study.

> *Expert analysis by the Commission's staff seismologist, geologists, and petroleum engineers concluded these results do not indicate any bounding faults in the immediate vicinity of the wells tested. At this time, there is no conclusive evidence the disposal wells tested were a causal factor in the May 7 seismic event. The tests were conducted to help determine the effect of injection operations on pressures within subsurface rock formations.*

Discussions of what can and cannot be determined from a falloff test are outside the realm of this book. Suffice it to say, however, that falloff tests **could not possibly** have determined the potential for earthquakes being caused by wastewater injection. And even though the data that the Commission used was released to the public, sufficient discussions of the engineering basis for their conclusions were not provided.

The lack of apparent technical rigor on this topic would seem imprudent for an important public safety issue, and certainly does not engender confidence in the Commission's ability to address complex technical issues. With the deep bench of oil and gas expertise available in the State, the Commission should be embarrassed. Is it any wonder that Texas voters believe that their Railroad Commission is more politically than technically competent?

Also in 2015, just-cause hearings were held by the Commission to determine whether the two wells identified by the SMU study should be found to have caused the Azle earthquakes. In spite of the findings from the SMU study, Commission staff concluded that there was no such connection. Hopefully the three Commissioners will override the conclusions of their staff and come to a more reasonable conclusion. If not, they will simply have proven that the regulations promulgated in 2014 were a farce.

It is impossible to establish a near-certain criterion that links seismic activity with specific wastewater injection wells. This means that developing an appropriate regulatory response will be difficult and fraught with uncertainty. With that said, though, here are five things that can be done to begin the process.

- **Increase the reporting requirements for wastewater injection wells.** Current Railroad Commission rules only require collecting wastewater well pressures and rates on a monthly basis, with public reporting only once a year. Given the public safety nature of possible seismic activity, transparency would dictate that rates and pressures be publicly reported more frequently. Producing oil and gas wells already report monthly data. Putting wastewater wells on the same schedule would not seem to be particularly burdensome.

- **Provide for auditable collection of daily injection rates and pressures for wells near seismic events.** Daily rates and pressures are now routinely captured by well operators. Having this data available to the public and to the Railroad Commission would provide valuable information for Texans who wish to evaluate the source of seismic activity near their homes.

- **Make water disposal wells subject to immediate curtailment when there is sufficient evidence of a link with earthquakes.** If wells are **actually** curtailed (rate either reduced or well shut in, even for a short time) by the Railroad Commission, wastewater well operators will be far more diligent in locating and operating their wells, knowing the financial risks they could potentially incur.

- **Provide a liability system for earthquake damage.** This could be done via posting of bonds or some other insurance-based solution, perhaps through a joint industry initiative. Homeowners who experience earthquake damage need to have some reasonable recourse when operations are likely responsible, even in the absence of negligence.

- **Increase monitoring of seismic events, along with studies to establish the scientific relationships between subsurface injection and earthquakes.** During the 84th Session, the Texas Legislature authorized $4.5 million for increased seismic data gathering and studies to investigate the causal links between wastewater injection and earthquakes. This seems like a modest investment for a clear public safety issue. It would have been preferable for such a study to have been funded by companies with commercial interests in wastewater disposal wells. However, given the level of public distrust of an industry-funded study, perhaps it is appropriate for it to be funded and conducted by the State.

With curtailment and liability procedures in place, wastewater disposal well operators would have the necessary incentives to minimize earthquake risk from wastewater wells. Increased transparency (in the form of well data) along with results from a publicly-funded study could go a long way toward helping mitigate future problems. We should fully expect that wastewater disposal well operators would move to appropriately locate wells (geologically) and to operate them at rates and pressures so as to maintain a low probability of seismic activity.

Some will argue that mandated data gathering is a form of command-and-control regulation. In a sense, they are correct. What is slightly different about data requirements is that it is mandating transparency as opposed to mandating specific operational actions. Since well data is already routinely measured at very modest cost, intrusive requirements in the name of transparency would seem to be justified. When commercial operations have impacts in the public sphere, the public legitimately should expect to have information available that is necessary for making assessments of the impact of those operations on their community.

Common carrier pipelines

The right of government to take private land for public use is enshrined in the *US Constitution's Fifth Amendment*:

> ... nor shall private property be taken for public use, without just compensation.

A similar provision exists in the *Texas Constitution*.

Unfortunately, the US Supreme Court in a narrow (5-4) decision of the Kelo v. City of New London case held that eminent domain can be used to transfer land from one **private** owner to another, reasoning that the public benefits that a community enjoys from economic growth qualify as an allowable "public use".

In a similar vein, Texas law allows eminent domain transfers by common-carrier pipelines to obtain rights-of-way. Texas statute provides that a pipeline company is a common carrier if it owns or operates a pipeline that transports product purchased from others, or transports product for hire. Pipelines that transport **only** their own product are not considered common carriers.

On its T-4 Application for Permit to Operate a Pipeline in Texas, the Railroad Commission requires a declaration as to whether a pipeline is private or a common carrier. However, the Railroad Commission has no statutory authority to either approve or disapprove common carrier status. In 2014, regulations were put in place by the Commission to provide additional transparency. These regulations will likely prove to be inadequately transparent. There have been calls for the Texas Railroad Commission to be more diligent in its determination of common carrier pipeline status.

Common carrier pipeline issues were most recently addressed by the Texas Supreme Court in the Texas Rice Land Partners Ltd. v.

Denbury Green Pipeline-Texas LLC case decided in 2011. On its form T-4 permit application to the Railroad Commission, Denbury Green Pipeline checked the box indicating that their pipeline was a common carrier. Denbury used this fact to obtain an injunction to seize land from Texas Rice Partners by eminent domain.

The landowner whose land was seized argued that the Denbury pipeline did not, in fact, qualify for common carrier status. The landowner ultimately prevailed before the Texas Supreme Court. Key findings from this case that are significant to this discussion include:

- Both the District Court and the Court of Appeals had previously affirmed, based on a variety of evidence, that Denbury Green **did** qualify as a common carrier.
- The Texas Supreme Court unanimously ruled that a Railroad Commission form T-4 permit alone does not establish a pipeline as a common carrier, and that the Denbury pipeline was **not** a common carrier.

Both of these facts suggest that even if the Railroad Commission had more extensively reviewed Denbury's application, it might have erred in much the same way as the lower courts did. Besides, the issue of common carrier status would still remain, regardless of the Railroad Commission's form T-4.

I (along with four US Supreme Court Justices) strongly disagree with the reasoning that transferring private property to a private entity for improved economic growth should qualify as public use. The problem with this concept of public use is that it can be used to justify almost **any** private transfer of property. Besides, when property is transferred to a private owner, clearly that owner derives significantly more benefit from the transfer than does the public at large.

With that caveat, the question is how Texas should move forward. Clearly, the Texas Railroad Commission has little or no current authority to either approve or deny common carrier pipeline status. Without too much concern about what powers the Commission has in this regard, the following are recommendations for the Texas Legislature to consider:

- **Transfer all pipeline regulation to the Texas Department of Transportation.** TxDOT already is deeply engaged in regulating rights-of-way for roadways and other monopolistic transportation activities. Removing these issues from Railroad Commission jurisdiction would allow the Commission to focus on is primary role of regulating subsurface oil and gas activities.

- **Improved diligence on common carrier status.** Commercial entities currently only need to declare themselves common carriers to be classified as such. When sued, of course, they are required to support that classification. Since common carriers have special status granted to them by the State, it is appropriate that they be required to be **more** transparent than other commercial entities. Transparency should come in the form of regularly reporting the fraction of pipeline capacity that is used to transport third-party product (*i.e.*, not owned by the pipeline operator).

- **Increase minimum third-party product fraction and other requirements in order for a pipeline to qualify for common carrier status.** Ideally, common carriers should **only** carry third-party product. However, requiring minimum third-party product fractions of at least 50% would not seem unreasonable. Given the seriousness of involuntary property seizure, there may well be other requirements that should be considered before granting a pipeline common-carrier status.

- **Harsh penalties for negligently claiming common carrier status.** Falsely claiming common carrier status when property is seized by eminent domain is theft. Restitution would be the appropriate remedy. The pipeline operator should be required to either: a) agree to appropriate restitution payments for land that was falsely seized, or b) restore any seized land to its original condition including ceasing a pipeline's operation. In either case, the subject pipeline's permit should be revoked.

- **Provisions for relinquishing common carrier status.** Should a common carrier pipeline later become private by ceasing to carry third-party product or by dropping below a certain capacity threshold, there should be provisions for relinquishing common carrier status. Though this action would not amount to theft, it should require compensation to landowners whose land was previously seized.
- **Additional provisions to make land seizure a last resort.** Land being seized by eminent domain should only occur when a carrier has been unable to arrive at a suitable mutual agreement with a land-owner. Ways to "level the playing field" in negotiations could conceivably include: a) requiring pipeline operators to first get an appraisal of the value of the **actual** section of land to be seized, b) setting the seizure price at 150% of the appraised price (based on the observation that a **willing** owner would sell at 100% of the appraised price), and c) requiring pipeline operators to pay legal fees if a seizure price is successfully challenged (though not vice-versa).
- **Provisions to encourage pipeline operators to use existing rights-of-way, including those along highways and existing pipeline routes.** Even widening rights-of-way on land that has already been seized would be less detrimental to landowners than seizing more pristine land.

Making it more difficult to take someone's property seems only just, given that the property is being seized by a private entity. Though some pipeline operators will undoubtedly object to these recommendations as being too harsh or cumbersome, they really are a small price to pay to protect the rights of property owners.

Groundwater contamination

Fear of contamination of groundwaters from fracking is a commonly-heard justification to halt the practice. Even though the US Environmental Protection Agency recently issued a report denying any serious contamination problem, many still have grave concerns about their drinking water.

Though contamination of groundwater by chemical-laden fracking fluids is most commonly talked about, another issue in the public eye is groundwater that has been contaminated by natural gas (methane). A water well in the Silverado on the Brazos neighborhood in Parker County owned by a Mr. Steven Lipsky provides a good example of the type of problem people fear.

Mr. Lipsky's well first began experiencing high levels of methane gas in 2010. Eventually the amount of gas became so large that Lipsky was unable to use his well, and subsequently was even able to light it on fire. Lipsky believed that the source of the methane was from nearby Barnett shale wells drilled by Range Resources.

The Texas Railroad Commission did a number of studies of the Lipsky well, culminating in a final report in May 2014 that concluded that the Range wells were **not** responsible for the gas. Range Resources has since sued Lipsky for public defamation. Since publication of the Commission's final report there has been considerable controversy, including reports that some scientists doubt the validity of the Railroad Commission's conclusions.

Those of us who are reasonable and open-minded (particularly when it comes to complex technical issues) are not well served by the he-said-she-said game that tends to be driven by those with axes to grind. So let's try to shed a bit of light on this subject, starting out with a brief discussion of the technical issues.

First, it is not uncommon for natural gas to exist either dissolved in or adjacent to groundwater zones. Two different processes are generally considered to produce hydrocarbon gases in the earth: *biogenic* and *thermogenic*. Biogenic processes occur at low temperatures largely through bacterial decomposition of organic material (*e.g.*, swamp gas, manure gas). Thermogenic gas is formed at deeper depths as a consequence of elevated pressure and temperature for long periods of time. Natural gas is produced in source rocks by thermogenic processes.

Without getting into too much geochemistry, suffice it to say that biogenic gas can be distinguished from thermogenic gas on the basis of composition. Thermogenic natural gases typically contain large amounts of compounds in addition to methane. Also, carbon atoms in thermogenic gases have an extra neutron in them. Different gas sources (*e.g.*, geologic layers) will also have different compositions and isotopic ratios.

It is difficult to know the true amount of methane that's actually **dissolved** in groundwater as opposed to being mobile as a separate gas phase. Unless water sampling is done downhole under pressure, dissolved methane will be released upon exposing the water sample to lower pressures at the surface.

In addition, some of the gas in the groundwater zone may exist as a *free gas* phase, totally separate from the water phase. The methane gas produced from a water well at the surface will at least have been partially separated from the water downhole, making an analysis of the total amount of gas that has migrated to a particular well difficult to discern.

Even when it has been determined that groundwater gas is of thermogenic origin (which it appears to be in the Lipsky case) there's the problem of attempting to determine the exact source of that gas, meaning which well(s), which zone(s), and when in time.

Obviously this is a complicated problem. Add to this the difficulty of inferring the state of things beneath the earth's surface where data is always sparse and largely unreliable. Much must be inferred and estimated to arrive at reasonable, often speculative, conclusions. Uncertainty about such matters is typically dealt with through multi-disciplinary data, complex analyses, and a certain level of speculative reasoning. In this case, studies would need to be conducted using geochemistry to determine possible stratigraphic sources of the gas, along with geology, petrophysical analysis, and fluid flow calculations to determine the location of potential migration paths.

With all that said, the Railroad Commission's report on the Lipsky well is (at least to this expert) technically unconvincing. In particular, it fails to adequately convince the claimant, not to mention the Texas public, about the adequacy of its analysis. The real question – where the gas in the homeowner's well is coming from – remains unanswered.

If you're an oil and gas industry supporter, you will likely believe Range Resources. If you're someone adamantly opposed to hydraulic fracturing, you probably believe Lipsky. Both of these views come from beliefs, not facts. What could have and should have been done in this case?

Lipsky's primary problem is that he does not have the ability or the resources to refute findings by either Range Resources or the Texas Railroad Commission. He has neither the data nor the expertise nor the money to be able to conclusively discern the likelihood that his well was contaminated by a Range well.

The issue here is that Lipsky and Range, as well as the mineral rights owner of Range's well, are both private entities. Perhaps in a more perfect world, private matters could and should be resolved privately. But in the world we actually live in, the power discrepancy between Range and Lipsky is too great. Normally, when private matters cannot be resolved privately, our judicial system is called upon to provide recourse and remedies. When there are highly-technical disputes, and with the cumbersome and expensive court system currently available to people like Lipsky, normal judicial proceedings seem futile.

Interestingly, the Railroad Commission has already established a system of quasi-judicial proceedings. Its hearings involve both a legal and a technical examiner, and in many ways are run like a courtroom (though a bit more expeditiously). In Lipsky's case, the Railroad Commission actually **did** gather data and **did** conduct studies, as unconvincing as they were. So what could have and should have been done differently? Some possibilities:

- **Utilize Texas' deep professional oil and gas bench.** As the oil and gas industry has become more and more technically sophisticated, it has become increasingly difficult for the Commission to stay abreast of rapid technological changes. This action would require budgeting for outside consultants in lieu of internal personnel. Though it's possible that the cost to the State could be higher, in my experience the use of highly-qualified experts often renders better results with smaller budgets.
- **Submit Commission reports to a peer-review process.** This is done routinely in the engineering and scientific communities on research studies that are far less important than the technical issues that come before the Commission. The "deep bench" that is used to perform studies could well become a solid peer review system overseen by Commission staff.

- **Allow hearings to be conducted by video conference.** This recommendation was made by the Sunset Commission, and is a good one. Claimants in remote locations (most oil and gas is nowhere near Austin) could easily be engaged in hearings from their home locations using many of the free or inexpensive means of communication available to the public. The Railroad Commission already has a number of field offices that could be used for claimants who did not have adequate video conferencing technologies available.

- **Provide public-defender type counsel available to claimants with small resources.** It would be disadvantageous to make this a large expenditure of Commission resources. The expedited nature of the Commission's proceedings would suggest that allocated resources would be small and worthwhile.

Improvements such as these would enhance the Commission's ability to provide adequate technical input into a streamlined quasi-judicial process that would undoubtedly better protect individual landowners in disputes such as the one between Lipsky and Range Resources.

An obvious pushback to these recommendations is in dealing with intentional claims of contamination when none actually occurred, such as was done for the 2010 Gasland documentary. Quasi-judicial proceedings, even with all their warts, are probably better at dealing with such issues than the current situation. By better engaging the large cadre of oil and gas professionals in the State, the Texas Railroad Commission might just be seen as increasingly trustworthy by the public at large.

Infrastructure

One of the more serious side effects of rapid oil and gas development has been with regard to infrastructure, particularly roads. Much of the activity in the most recent boom was in relatively sparsely-populated counties with a small number of roads and small highway budgets. But even in larger counties impacted by oil and gas, the strain on roadways can be significant.

Ideally, companies utilizing the roadways should pay for their share of maintenance and reconstruction. This, in fact, is being done in some municipalities. Some even require road damage bonds be posted by oil and gas operators. But in many counties and other entities responsible for road construction and maintenance, funding has not always been available to keep up.

One of the basic problems with road construction in Texas is that while local counties are responsible for many of the roads, funding often flows from the State and is not under the control of local officials. A number of proposals have been put forward to remedy this problem, but all have failed to pass the Texas legislature. One of the more interesting proposals is to allow counties and municipalities to collect royalties from oil and gas beneath the roadways for which they are responsible.

Proposition 1, passed by Texas voters in 2014, was enacted to help solve some of Texas' roadway needs – both those related to population growth as well as oil and gas operations. Proposition 1 altered a previous constitutional amendment that, in 1988, established the Economic Stabilization (Rainy Day) Fund. Prior to Proposition 1, this fund received 75% of oil and gas production severance taxes in excess of those collected in 1987.

Proposition 1 directed half of the money that was earmarked for the Rainy Day Fund to be transferred to the State Highway Fund, but only to be used for constructing, maintaining, and acquiring rights of way for public roadways other than toll roads. One of the positive aspects of Proposition 1 is that severance taxes effectively are being used to help support highway deterioration resulting from oil and gas operations.

Additional local control over roads is still needed to more effectively deal with the infrastructure needs unique to each different locality. The following are two ideas that are worth considering:

- **Allow counties and local entities to collect royalty revenues from underneath surface property being maintained by them.** Even if the State collects royalty payments, a simple pass-through would eliminate the pull-and-tug that occurs in state agencies and the Legislature in re-allocating resources to local entities.
- **Block-grant a fixed fraction of locally-collected severance taxes back to local taxing authorities.** Even a portion of the 25% "excess" severance taxes that are currently placed in the general revenue fund would be very beneficial in helping local entities deal with their infrastructure needs that result from oil and gas operations.

To avoid funding going to the largest counties with the most well-connected constituents, this approach would ensure that the counties with significant impacts from oil and gas activity would receive funds more commensurate with those activities. Too often, top-down organizations (the State of Texas is not immune) delegate local responsibility but retain central control. This is a recipe for frustration by voters, who look to their local representatives to fix local problems, even when those representatives are hamstrung by resources that come down from on-high with all the usual strings attached.

Local fracking bans

In November 2014 an initiative was passed by voters in the City of Denton that banned hydraulic fracturing within city limits:

> *SHALL AN ORDINANCE BE ENACTED PROHIBITING, WITHIN THE CORPORATE LIMITS OF THE CITY OF*

> *DENTON, TEXAS, HYDRAULIC FRACTURING, A WELL STIMULATION PROCESS INVOLVING THE USE OF WATER, SAND AND/OR CHEMICAL ADDITIVES PUMPED UNDER HIGH PRESSURE TO FRACTURE SUBSURFACE NON-POROUS ROCK FORMATIONS SUCH AS SHALE TO IMPROVE THE FLOW OF NATURAL GAS, OIL, OR OTHER HYDROCARBONS INTO THE WELL, WITH SUBSEQUENT HIGH RATE, EXTENDED FLOWBACK TO EXPEL FRACTURE FLUIDS AND SOLIDS.*

This ordinance, had it stood, would have effectively vetoed the State of Texas' established preemption over the regulation of subsurface oil and gas operations. The response to this initiative should have been expected. Immediately following the November election, a lawsuit was filed by the Texas General Land Office and the Texas Oil and Gas Association. Even before the suit could be heard, the Texas Legislature set to work. A number of bills were filed by the 84th Legislature that would have put major constraints on the ability of voters in Texas home rule cities to engage in initiatives and referendums. Some could have effectively shut down local initiatives altogether.

One such bill did pass the Texas Legislature and was signed by Governor Abbott – House Bill 40. HB 40 was intended to clarify where local control over oil and gas operations ends and Texas law begins. Such clarification was already established by Texas law and legal precedent. HB 40 went further than in the past.

HB 40 expressly preempts local authorities from regulating **subsurface** oil and gas operations, though continues to allow local ordinances that regulate **only** aboveground oil and gas activities. These include regulations governing fire and emergency response, traffic, lights, or noise, or imposing notice or reasonable setback requirements. This is effectively where Texas law stood prior to passing HB 40.

Where the bill extended State jurisdiction was in preventing local authorities from disallowing oil and gas operations that were "commercially reasonable", and specifying that local ordinances must not "effectively prohibit an oil and gas operation conducted by a reasonably prudent operator".

HB 40 defined "commercially reasonable" as a condition that would allow an operator to economically conduct oil and gas operations, based on the objective standard of a reasonably prudent operator and not on an individualized assessment of an actual operator's capacity. It is hard to imagine how this latter language will not lead to a raft of lawsuits.

A 2014 paper by David Spence, Professor of Law, Politics, and Regulation at the University of Texas at Austin School of Law and McCombs School of Business, provides an excellent discussion of the issues surrounding local vetoes such as the one attempted by the City of Denton. Professor Spence points out that the fundamental issue centers around the fact that local entities have a disproportionate share of the risks of oil and gas development, both perceived and real, compared to their share of the economic benefits.

State preemption over local oil and gas ordinances is very much parallel to another issue affecting surface rights owners. In Texas it is common for mineral rights to be severed from surface property rights. Since under Texas law, surface rights are subservient to mineral rights, surface owners are increasingly impacted, with little say over or financial compensation for, operations near or sometimes even on their property. Not only is there an issue of preemption, but mineral rights owners might actually have had takings claims that required compensation from the City of Denton. Takings claims are enshrined in both the Texas and US constitutions.

Major players in the oil and gas industry appear to be aware of the impact that their operations are increasingly having on surface rights owners throughout Texas and the rest of the US. The American Petroleum Institute publishes a set of Community Engagement Guidelines that seeks to address some of the issues. Though the document is short on details, it clearly indicates that the oil and gas industry is aware that operators should be actively engaged with local communities as their activities expand into inhabited areas.

Denton's issue was essentially a conflict between mineral owners, who have the right to realize the economic value of their property, and surface owners, who have the right to the quiet enjoyment of their property. Resolving the dilemma of these conflicting rights involves difficult choices. These choices are best resolved, where possible, either privately or at the local level. Local governments, after all, are closest and most responsive to local voters.

Geographically expansive shale development means that increasing numbers of surface property owners are being impacted by oil and gas activities, but without the economic levers and rewards that landowners had when oil and gas operations were more rural and when mineral and surface rights were coincident. As shale exploitation continues in the State, and it will, as Texas towns and cities grow, and they will, this problem will only get worse.

The following are three things that should be considered that would help provide landowners and local entities with the ability to enable responsible oil and gas operations while still protecting the rights of citizens to the quiet enjoyment of their homes and property.

- **Block grant a fraction of severance taxes collected back to local jurisdictions.** These funds would provide additional resources for local entities to help deal with the negative repercussions of oil and gas surface operations. Additionally, this action would provide financial incentives for local entities to make reasonable accommodations for both surface and mineral owners.

- **Change Texas statutes to better balance the rights of surface and mineral owners by encouraging mutually-agreeable settlements.** The total dominance of mineral rights over surface rights unnecessarily establishes economic rights as being more important than quiet enjoyment rights. Possible negotiated remedies could include:

    » Royalty sharing between surface and mineral owners

    » Per diem payments to surface owners during drilling and completion operations

    » Buyouts of surface property

    » Buyouts of subsurface mineral rights

- **Use the quasi-judicial role of the Railroad Commission to resolve surface and mineral rights conflicts.** Historically, the Commission has not been involved in these disputes. But much as the Commission was able to establish remedies for balancing the rights of mineral owners and operating companies, it is clear that the Commission could play an important role in developing clear statutes and regulations that would benefit many Texans.

By putting surface rights on a more equal footing with mineral rights, and by providing more local oversight over the resolution of conflicts between the two, local governing agencies will be in a better position to find local solutions that best suit the needs of communities as well as mineral owners.

The Texas way

Many Texans chafe at intrusive government regulations. We no more like state government dictating our lives than we do the federal government. But in the world we currently live it, a certain level of regulation may be necessary to protect our property and our liberties.

Most of us would prefer to have the personal ability and power to protect our own self-interests. When our self-interests come into conflict, however, with large and powerful non-governmental entities, there may be a role for government to provide appropriate mechanisms for adjudication.

Texans are fiercely independent doers and problem solvers. A bottom-up system that empowers Texans is far better than a top-down system that disempowers us. Regulatory agencies such as the Texas Railroad Commission could and should be transformed into protectors of individual rights and property. Such a cultural shift, however, will not occur organically. It will require increased voter awareness to change this important agency for the better.

FURTHER READING

Below is a list of important resources used in the writing of this book.

American Petroleum Institute: **Community Engagement Guidelines**, ANSI/API Bulletin 100-3, First Edition, July 2014, http://www.api.org/~/media/Files/Policy/Exploration/100-3_e1.pdf

Childs, W.R.: **The Texas Railroad Commission: Understanding Regulation in American to the Mid-Twentieth Century**, Texas A&M University Press, College Station, TX (2005).

Clark, N.J.: **Elements of Petroleum Reservoirs**, Society of Petroleum Engineers, Richardson, TX (1960).

Drollette, B.D., K. Hoelzer, N.R. Warner, T.H.Darah, O. Karatum, M.P. O'Conner, R.K. Nelson, L.A. Fernandez, C.M. Reddy, A. Vengosh, R.B. Jackson, M. Elsner, and D.L. Plata: "Elevated Levels of Diesel Range Organic Compounds in Groundwater Near Marcellus Gas Operations are Derived from Surface Activities," *Proceedings of the US National Academy of Sciences*, 16 Sep 2015, http://www.pnas.org/content/early/2015/10/07/1511474112.abstract

Durrett, B.: "A Primer on Oil and Gas Regulation in Texas: Spacing, Density, Permits, Exceptions," **Landman**, Winter 2013-14, http://www.burlesonllp.com/D6B628/assets/files/Documents/Durrett_Pub-NA.pdf

Engineering Toolbox: "Energy Content in Common Energy Sources," http://www.engineeringtoolbox.com/energy-content-d_868.html

Gold, R.: **The Boom: How Fracking Ignited the American Energy Revolution and Changed the World**, Simon & Shuster, New York (2014).

Hornbach, M.J., H.R. DeShon, W.L. Ellsworth, B.W. Stump, C. Hayward. C. Frohlich, H.R. Oldham, J.E. Olson, M.B. Magnani, C. Brokaw, and J.H. Luetgert: "Causal Factors for Seismicity Near Azle, Texas," **Nature Communications, 6**, Article No. 6728, 21 Apr 2015, http://www.nature.com/ncomms/2015/150421/ncomms7728/full/ncomms7728.html

Kondash, A. and A. Vengosh: "Water Footprint of Hydraulic Fracturing," **Environment Science & Technology Letters**, 31 Aug 2015, http://pubs.acs.org/doi/pdf/10.1021/acs.estlett.5b00211

Murray, C.: **By the People: Rebuilding Liberty Without Permission**, Crown Forum, New York (2015).

Prindle, D.F.: **Petroleum Politics and the Texas Railroad Commission**, University of Texas Press, Austin (1981).

Quaschning, V.: "Renewable Energy and Climate Change: Specific Carbon Dioxide Emissions of Various Fuels)," http://www.volker-quaschning.de/datserv/CO2-spez/index_e.php

Petroleum Extension Service: **Fundamentals of Petroleum**, 3rd edition, M. Gerding (ed.), U. of Texas, Austin, TX (1986).

Railroad of Commission of Texas: http://www.rrc.state.tx.us/

Railroad Commission of Texas: **Water Well Complaint Investigation Report, Silverado on the Brazos Neighborhood, Parker County Texas**, 23 May 2014,
https://www.documentcloud.org/documents/1175488-water-well-complaint-investigation-report-5-23.html

Rochlin, G.I., T.R. La Porte, and K.H. Roberts: "The Self-Designing High-Reliability Organization: Aircraft Carrier Flight Operations at Sea," **Naval War College Review**, Autumn 1987, http://fas.org/man/dod-101/sys/ship/docs/art7su98.htm

Spence, D.B.: "The Political Economy of Local Vetoes," The University of Texas School of Law," **Public Law and Legal Theory Research Paper Series Number 552**, 02 Mar 2014, http://papers.ssrn.com/sol3/papers.cfm?abstract_id=2403978

Texas State Historical Association: "East Texas Oilfield," https://tshaonline.org/handbook/online/articles/doe01

U.S. Energy Information Administration: "Average Price of Electricity to Ultimate Customers by End-Use Sector," http://www.eia.gov/electricity/monthly/epm_table_grapher.cfm?t=epmt_5_6_a

U.S. Energy Information Administration: "How much carbon dioxide is produced when different fuels are burned?," http://www.eia.gov/tools/faqs/faq.cfm?id=73&t=11

U.S. Energy Information Administration: "How much coal, natural gas, or petroleum is used to generate a kilowatt-hour of electricity?," http://www.eia.gov/tools/faqs/faq.cfm?id=667&t=2

U.S. Energy Information Administration: "How much electricity does an American home use?" http://www.eia.gov/tools/faqs/faq.cfm?id=97&t=3

Whitworth, H.P.: "Land and Regulatory Issues Related to Horizontal Wells," **Texas Mineral Title Course**, Institute for Energy Law, The Center for American and International Law, Houston, TX, 02-03 May 2013, http://www.cailaw.org/media/files/IEL/ConferenceMaterial/2013/TexasMineralTitle/FWithworth-paper.pdf

Wikipedia: “Dunbar’s number,”
https://en.wikipedia.org/wiki/Dunbar%27s_number

Wikipedia: “East Texas Oil Field,”
https://en.wikipedia.org/wiki/East_Texas_Oil_Field

Wikipedia: “Energy Density,”
https://en.wikipedia.org/wiki/Energy_density

Wikipedia: “Peak Oil,”
https://en.wikipedia.org/wiki/Peak_oil

ABOUT THE AUTHOR

After graduating with a BS in Engineering from Harvey Mudd College in 1972, Dr. Miller began a career in the oil and gas industry as a petroleum engineer. Mark later went on to earn a PhD in petroleum engineering from Stanford University prior to an 18-year teaching stint at The University of Texas at Austin. After abandoning university life, Mark established a worldwide petroleum engineering consulting practice, and was a founder and CEO/CTO of a small company that provided software to the oil and gas industry. He is currently retired and does occasional consulting. Mark was the 2014 Libertarian Party nominee for Texas Railroad Commissioner.

Mark is married and has two sons and two grandsons living in Austin, Texas.

Made in the USA
San Bernardino, CA
22 August 2016